BLACK AND WHITE IDENTITY FORMATION

WILEY SERIES ON PSYCHOLOGICAL DISORDERS

IRVING B. WEINER, Editor

School of Medicine and Dentistry
The University of Rochester

BLACK AND WHITE
IDENTITY FORMATION

Studies in the Psychosocial Development
of Lower Socioeconomic Class Adolescent Boys

STUART T. HAUSER, A.M., M.D.

Wiley-Interscience

A DIVISION OF JOHN WILEY & SONS, INC.

NEW YORK • LONDON • SYDNEY • TORONTO

Library of Congress Catalog Card Number: 77-138910

ISBN 471 36150 X

Printed in the United States of America

10 9 8 7 6 5 4 3 2 1

To Barbara

Series Preface

This series of books is addressed to behavioral scientists concerned with understanding and ameliorating psychological disorders. Its scope should prove pertinent to clinicians and their students in psychology, psychiatry, social work, and other disciplines that deal with problems of human behavior as well as to theoreticians and researchers studying these problems. Although many facets of behavioral science have relevance to psychological disorder, the series concentrates on the three core clinical areas of psychopathology, personality assessment, and psychotherapy.

Each of these clinical areas can be discussed in terms of theoretical foundations that identify directions for further development, empirical data that summarize current knowledge, and practical applications that guide the clinician in his work with patients. The books in this series present scholarly integrations of such theoretical, empirical, and practical approaches to clinical concerns. Some pursue the implications of research findings for the validity of alternative theoretical frameworks or for the utility of various modes of clinical practice; others consider the implication of certain conceptual models for lines of research or for the elaboration of clinical methods; and others encompass a wide range of theoretical, research, and practical issues as they pertain to a specific psychological disturbance, assessment technique, or treatment modality.

IRVING B. WEINER

University of Rochester
Rochester, New York

Preface

This monograph presents, discusses, and offers approaches to understanding the results of longitudinal studies of white and black adolescents. Multiple facets of method, data, and theoretical models are in turn taken up within the following chapters. However, two important issues are not directly addressed within the text, although they are alluded to in many places. These issues touch upon the interpretation and implications of our findings. In these preliminary notes, I raise the problems and thereby encourage the reader to keep them in mind as he then proceeds through the more detailed theory, method, and data.

The issues are related to empirical concerns. First of all, there is the matter of the *nature* of the sample. The group of adolescents that were longitudinally studied was a small one. It was of sufficient size to permit meaningful application of nonparametric statistical tests. The results of these tests, graphical techniques, and clinical approaches certainly argue for meaningful differences *within* the sample. At the same time, it is crucial to remember that these are findings derived from a small matched interracial sample. They are *not* results that in any way, by themselves, typify a "black ghetto" or "white slum." Such a leap in reasoning is ever tempting, and always fallacious. Nonetheless, even given this note of caution we are left with a serious problem; namely, how are we to understand findings that show the black adolescents to have different identity formation patterns from white adolescents. Much of the monograph is devoted to considering this extremely fascinating and perplexing question (Chapters 5–7).

A second issue is the era within which this study was carried out. These adolescents were followed between 1962 and 1967. The many events since then—ranging from major civil rights demonstrations to new organizations and assassinations—may have generated significant impacts upon those aspects of the sociocultural surroundings important for adolescent development. Thus contemporary white and black adolescents might show very different identity patterns from those within our sample. This is an important kind of speculation, for it is directed toward attempting to understand the developmental impacts of these seemingly major societal events.*

To complete a longitudinal study of this kind required the support and cooperation of many people, most of all the subjects themselves. Their names have been disguised within the text for the sake of confidentiality. This does not detract from the debt I think the investigation owes to them for their sustained cooperation in allowing us to observe them growing up.

A number of other people made significant contributions to the planning, carrying out, and completion of this investigation. Much of the original planning and testing was done while I was a medical student at Yale University School of Medicine. My advisor in these phases was Dr. Ernst Prelinger. His teaching, guidance, and support was crucial in enabling the study to be conceived and then taken through the first critical stages of data collection and analysis.

In the later stages of the work, the further analysis and understanding of results, Dr. Elliot Mishler was most instrumental. Our many discussions, his careful reading of the manuscript, his many suggestions about analysis and display of data represent but a portion of his contribution. Indeed, his intellectual and personal support of my work is most directly responsible for this monograph.

While working with Dr. Mishler in the social psychiatry laboratory of Massachusetts Mental Health Center, a number of other members of the laboratory were extremely helpful. Dr. Nancy Waxler read early portions of the manuscript and offered valuable suggestions concerning statistical techniques and tables. Members of the Research Training in Social Psychiatry seminar were helpful in discussing and questioning aspects of the study; I am grateful for their stimulating questions and interest to H. Stephen Leff, Kent Ravenscroft, Virginia Abernathy, Eve Bingham, and Ron Walter.

I am, in addition, greatly indebted to Dr. Jack Ewalt, Superintendent of Massachusetts Mental Health Center and Professor of Psychiatry at Harvard Medical School, for his continued encouragement and support of this research over the last two and one-half years.

More specifically I am indebted to Drs. Kenneth Keniston, Florence Shelton, Peter Reich, John Mack, and George Vailliant for several valuable discussions about facets of the study. In addition, the writings of Erik Erikson, David Riesman, and Lee Rainwater have influenced much of the design and interpretation of the investigation. The sensitive and intelligent comments of Dr. Irving B. Weiner, the series editor, were of much assistance in shaping the final form of the monograph.

---

* For this reason, a current identity and cognition study using a larger sample of white and black adolescents is now under way.

Finally, I want to acknowledge the help of several people in the many and tedious tasks involved in performing the studies. Charles Twyman, Grace Dowdy, Arnold Lerner, Gerald Barberisi, Nicholas Defeo, Hildegarde Schwartz, and Robert Schreck were most helpful to me in obtaining research subjects and neighborhood facilities in New Haven. Richard Cash and Susan Bender were helpful in early recording work.

Roger Bakeman, with his intelligent help in programming, allowed the data to be analyzed through the Yale Computer Center. Two research assistants, Kathleen McCormack and Gerry Pilgrim were most helpful through their stimulating questions and practical assistance in the last revisions of the manuscript. A series of secretaries have contributed in flexible and efficient ways to the several versions of the manuscript: Anne Granger, Judy Wettenstein, Jacqueline Hooker and, in the final important stages, Nancy Fuller. I am equally grateful to Linda Twing who typed the stencils for the final copy.

I am indebted to two sources for financial support of the project: in the summers of 1962–1965 Yale University School of Medicine provided National Institute of Health summer research fellowships. Foundations Fund for Research in Psychiatry provided a small grant to aid in the final collection and transcription of interview data.

Last, but certainly not least, I am greatly indebted to my wife, Barbara Hauser. In all phases of this project, she has helped directly with her intelligent discussions and editing. Even more important has been the less tangible aspect of her assistance; through her patience and support of the time and energy the project demanded of her husband. Her role in these many facets of the work is most appreciated and really reflected throughout the monograph.

STUART T. HAUSER

Boston, Massachusetts
June, 1970

Contents

CHAPTER 1

An Overview

THEORY

Professional as well as popular writings now reflect the "black revolution." Articles and studies of every conceivable kind flood bookshelf and newsstand, ranging from surveys of thousands of Negroes and whites to detailed investigations of specific historic, economic, and political ramifications of contemporary American Negro problems.[1] To be sure, the current trend includes psychological and psychiatric papers too, although their rate of appearance is somewhat slower than other kinds of investigations. Observations and tests of black children in the midst of school desegregation and social psychological perspectives on the Negro in America are among the more outstanding contributions in this genre.[2] In view of this plethora of writings one might ask, with some justification, "Why another such study?" Obviously, the investigation reported here also has roots in the difficult and perplexing current issues of American race relations. However, it differs in an important way from most of the other Negro studies. While begun before Pettigrew's plea for "theoretically guided" research on Negro problems, it was with this orientation in mind that the current study was designed and undertaken.[3]

In recent years there has been increasing interest in clarifying relationships between personality development and sociocultural context.[4] This

[1] See "The Negro in America," *Newsweek*, July 1963, and more recently, *Look*, January 7, 1969, for such surveys. More detailed studies are "The Negro American," *Daedalus*, 1965, and "The Negro Protest," *Annals of the American Academy of Political and Social Sciences*, 1965.

[2] See R. Coles, *Children of Crisis*, 1967; T. Pettigrew, *Profile of the American Negro*, 1964; and, more recently, *Black Rage*, 1968—the more clinical and social criticism by W. H. Grier and P. M. Cobb.

[3] T. F. Pettigrew, "Negro American Personality: Why Isn't More Known?," *Journal of Social Issues*, **20**, 1964.

[4] See B. Kaplan (Ed.), *Studying Personality Cross-Culturally*, 1961; C. Dubois,

study represents still another expression of such an interest in the question of the influences between context and psychological development. Careful and detailed studies of social and cultural systems have been present within the social science literature for many years. Similarly, although at times with somewhat less empirical rigor, psychoanalytic investigators have attempted systematic study of individual development, of personality systems, and of personality structures. A most difficult task has remained: the conceptualization and empirical study of the relationships between these understandings of psychological development and of social reality.

Early approaches to this interface problem were taken by Heinz Hartmann and his associates.[5] These works were then followed by those of Erik Erikson, who concentrated more specifically on the various phases of development, ranging from infancy to old age. Erikson's theory outlines a sequence of phases of psychosocial development and relates these phases to psychosexual development. Within each phase of the individual life cycle there is a phase-specific developmental task which must be solved by the growing individual in order to continue in his psychological maturation. Fundamental to this theory of development is its

". . . conceptual explanation of the individual's social development by tracing the unfolding of *the genetically social character* of the human individual in the course of his encounters with the social environment at each phase of his epigenesis."[6]

Of further importance is the theory's position that ". . . the society into which the individual is born makes him a member by influencing the *manner in which* he solves the tasks posed by each phase of his epigenetic development."[7] The phases of development are conceived of as universal. The typical solutions to the developmental tasks are conceived of as varying from society to society.

The People of Alor, 1944; E. Erikson, *Childhood and Society*, 1950; H. Hartmann, E. Kris, and P. M. Lowenstein, "Some Psychoanalytic Comments on Culture and Personality," in *Psychoanalysis and Culture*, G. Wilbur (Ed.), 1949. The above all represent the initial and continuing interest in such problems. More recent work is represented by such books as E. Erikson's *Young Man Luther*, 1958; *Identity: Youth and Crisis*, 1968; and *Ghandi's Truth*, 1969; K. Keniston's *The Uncommitted*, 1966; S. M. Elkin's *Slavery*, 1963; R. J. Lifton's *Thought Reform and the Psychology of Totalism*, 1961; and the anthology edited by R. W. White, *The Study of Lives*, 1963.

[5] H. Hartmann, *Ego Psychology and the Problem of Adaptation*, 1939 and Hartmann, Kris, and Lowenstein, *op. cit.*, are two of the best known statements of this early approach.

[6] D. Rapaport, "A Historical Survey of Psychoanalytic Ego Psychology," *Psychological Issues*, **1**, 15, 1959. Italics are Rapaport's.

[7] *Ibid*.

One well-known phase in development is adolescence. Long a major area of interest for students of human development, this period has become a prominent focus for social scientists as well. The literature of adolescence reflects the dual perspectives embodied in these two approaches:

"One [approach] lays its stress on the more individual aspect of the problem, as in psychoanalysis; the other regards the individual adolescent as a member of society, and considers his actions, his problems, even his aspirations to be the resultant of forces which have their origin in social developments."[8]

With the theoretical developments presented by Hartmann and Erikson, the potential dichotomy between these two approaches has lessened. Indeed, it is now clear that there are highly significant sociocultural aspects of individual adolescent development, and the grounds for this position are both theoretical and empirical.[9]

Continuing work in the direction of further careful understanding of the interface between sociocultural context and adolescent development is the specific orientation of this study. Concepts relevant to the investigation of adolescence from this perspective are those of *identity formation*. These are the notions with which Erikson formulates the basic psychosocial processes of adolescent development.

During adolescence problems of coherence, continuity, meaningfulness, and self-definition may and frequently do take precedence in individual awareness. At times these problems take on overwhelming importance. It is within this period of development that correspondences between childhood expectations and envisioned adulthood are then sought, and often not found.

The years preceding adolescence are ones of relative calm. Oedipal conflicts have for the moment subsided. Sexual impulses are, again for the moment, dormant. This is a time of much practical learning. The school-age child is learning skills which directly introduce him to the complex technology and thought of his surrounding society. Parents and teachers, both formal and informal, are seen increasingly as being among the crucial representatives of society. In other words, emergent at this point is a growing awareness of roles and relationships outside the immediate family setting.

With the advent of puberty, there occur major shifts in psychological as

[8] H. Deutsch, *Selected Problems of Adolescence*, 1967, p. 10.

[9] To some extent all of the works cited in footnote 4 present evidence for this position. It is discussed further by Deutsch. In addition, the recent anthology by E. Brody, *Minority Group Adolescents in the United States*, contains several pertinent papers.

well as physical development. Previously established ideals, rules, defenses, and perceived continuities are now rearranged or rejected.[10] In addition to the internal biological stresses of intensified sexual drives and rapidly advancing bodily changes, there are numerous pressures of a more social nature. These include issues of new physical and psychological intimacies. Present as well are those issues related to the social setting itself: the choices of occupational role, social standing within the local community, and the heightened presence of the peer group. Experienced is a seeming urgent pressure to make what appear to be irreversible commitments: commitments of a personal, sexual, occupational, and at times ideological nature. Within this turmoil those self-conscious and disquieting questions of adolescence become more conscious: "What do they think of me? Do they like me? What do I stand for? Where do I belong?" Definition, synthesis, and continuity are the leading themes of this turbulent time:

"What the regressing and growing, rebelling and maturing youth are now primarily concerned with is *who* and *what* they are in the eyes of the wider circle of significant people as compared to what they themselves have come to feel they are; and how to connect the dreams, idiosyncrasies, roles, and skills cultivated earlier with the occupational and social prototypes of the day.

. . . the ego of the adolescent is in great need of support, yet paradoxically it has to provide this support out of its own resources. Against newly intensified impulses it has to maintain the old defenses and create new ones; it has to consolidate achievements that have already been reached. The most important of its tasks is the struggle to synthesize all childhood identifications as they become enlarged and enriched by new ones. The successful end result of this struggle will be the formation of a solidified personality, endowed with a subjective feeling of identity that is confirmed and accepted as such by society."[11]

Here, then, are important and specific kinds of adolescent development problems. They are generated by an intertwining of intrapsychic and social forces. They are the problems of identity formation, the developmental issues to which this study is addressed.

Theoretical and empirical complexities are associated with the concepts of identity formation. Before we can study identity formation processes, such complexities must be both recognized and analyzed.

The conceptual framework underlying this work is that of psychoanalytic

[10] This general problem of adolescent shifts and "rearrangements" is discussed by A. Freud, *The Ego and Mechanisms of Defense*, 1948.

[11] E. Erikson, *Childhood and Society*, 1950, p. 307; and H. Deutsch, 1967, p. 33.

ego psychology. The overall perspectives and questions posed are to a large extent guided by psychoanalytic formulations about adolescence. Still more specifically, the choice of hypotheses and problems is based on the recent discussions of Erikson (1956, 1958, 1968).[12] In these latter clinical and theoretical formulations, the notion of identity formation is a basic one for Erikson's analysis of psychosocial issues, particularly those of adolescence. Identity formation represents the organization of a number of intrapsychic and psychosocial components, a synthesis encompassing a wide array of ego functions:

". . . an evolving configuration of constitutional givens, idiosyncratic libidinal needs, favored capacities, significant identifications, effective defenses, successful sublimations, and consistent roles."[13]

It is the formation of this configuration that is seen by Erikson as both a focus for and source of critical psychosocial problems in adolescence.

On several occasions Erikson (1950, 1958, 1959, 1968) discusses the theoretical implications and empirical referents of identity formation. Repeatedly, he and others working in this area have noted theoretical issues such as the relationship of identity to the older more systematized psychoanalytical concepts of ego, ego ideal, super-ego, identification, and self. Moreover, there is much to suggest numerous links between identity formation and nonpsychoanalytic concepts. How, for example, does one differentiate among the notions of ego identity, identity formation, role, self-system, persona, and life style? This is an especially important problem, for some of the issues referred to by identity formation and ego identity overlap with those of the social psychologist for whom role, self-systems, and other such terms are central ones. Clearly, if there is not to be duplication of effort and loss of needed comparative data, methods of "translation" or reconciliation of these concepts are needed. The problem is particularly apparent here, since with but rare exception all major psychological studies of Negroes have been made by social psychologists.[14]

In addition to these conceptual issues, there are a host of empirical questions raised by the notion of identity formation. In general, research into identity development has drawn from a very special minority group: white, middle socioeconomic class American youth (usually college students).[15] Such a sampling bias restricts legitimate generalizations about

[12] All dates in parentheses following an author's name refer to the corresponding entries in the bibliography.

[13] E. Erikson, "The Problem of Ego Identity," *Psychological Issues,* **1**, 116, 1959.

[14] Exceptions here include the earlier noted works of Coles and also papers by Poussiant (1968) and Pierce (1967); and most recently, Grier and Cobb (1968).

[15] See, for instance, G. B. Blaine and C. C. McArthur, *Emotional Problems of the*

identity processes as well as empirical testing of underlying assumptions. There are multiple questions that must be asked about identity formation in other social and cultural groups. Is the process a significant one in individuals of all populations? If it is, do identity development problems nonetheless take different forms and find very different solutions in other groups? Is identity development the predominant task of adolescence in all sociocultural groups? Or does identity formation emerge as a critical problem in other developmental periods for individuals of differing sociocultural contexts? None of these questions can satisfactorily be answered until identity studies draw their sample from a greater number and variety of human populations. Neither can the two fundamental assumptions implied by the concept of identity formation be validated. These assumptions were embodied in the preceding questions.

1. The formation of identity is a developmental problem found in all social and cultural groups. Though the process may be exacerbated or foreshortened, its occurrence is nevertheless not "culture-bound."

2. The formation of identity emerges as a critical psychosocial problem in late adolescence. Studies in the United States suggest that this period is in the late teens or early twenties.[16]

Specific formulations, and their implied predictions, about identity formation are detailed later. This survey of the issues raised by the concept also suggests the orientation taken by the following study. It is a study most avowedly intended as an exploration of identity development in varied sociocultural contexts. With this end in mind, lower socioeconomic

---

Student, 1961; B. M. Wedge, *Psychosocial Problems of College Men*, 1958; P. Blos, *On Adolescence*, 1962. Prelinger (1962) has suggested several reasons for this preferred sample: (1) identity development occurs over a relatively longer period of time for college students; (2) college students are confronted with both a wider range of possible career choices as well as social and ideological commitments; (3) college students usually have a greater variety of intellectual, social, and cultural resources available to them allowing "the development of more differentiated and various ego identities"; (4) college students are easily available for study; and (5) college students speak a language understood by identity investigators, who have relatively similar backgrounds. Studies by Erikson (1950, 1958), Lifton (1962), Ross (1962), and Prelinger (1958) represent exceptions to this bias.

16 It is of course difficult to define adolescence and late adolescence cross culturally. The start of the period might be viewed biologically as the onset of puberty, as the time of certain sexual and general bodily changes. However, what late adolescence means remains a problem. One possible solution might be to avoid the term "late adolescence" and speak instead of amount of time since puberty. This would emphasize the possibility that this period, and perhaps the other developmental ones as well, has a large degree of variation in extent and form within the limits imposed by certain biological constants.

class Negro and white boys were chosen as research subjects to be followed closely over a three-year period. Guiding the research was Erikson's description of identity formation, which has been given above in its most complicated and ambiguous form. Though possibly of clinical usefulness in this form, the description requires several changes in the direction of greater explicitness and empirical specification for nonclinical investigative purposes. The result of these changes is a more "operational" definition, which is discussed and presented in subsequent chapters.

Dignan proposes a definition of identity. She suggests ego identity is, "The complex of self-referent images which evolves through social interaction, thereby delimiting the self."[17] Although commendable for its relative clarity, her definition lacks emphasis on certain processes which are crucial to the notion of identity formation, namely, continuity and synthesis. Moreover, the unconscious aspects of identity are ignored by her definition. Sister Dignan is one of several students who have addressed themselves to the empirical study of identity formation. In addition to the obvious relevance of such specific identity studies, also pertinent to the research reported here are those works dealing with social class, the American Negro, and selected aspects of adolescent development. Aside from those of the social sciences and psychoanalysis, there are a number of relevant literary contributions. Following a selective review of this literature,[18] we return to the current study in order to deal further with the basic concepts and technical underpinnings.

RELATED PERSPECTIVES

Several of Erikson's original descriptions and clinical discussions of ego identity appear in *Childhood and Society* (1950), *Identity and The Life Cycle* (1959), and *Identity: Youth and Crisis* (1968). The latter collection includes papers rich in theoretical and clinical considerations, displaying as it were the immense complexity of both the identity construct and the events—psychosocial and intrapsychic—which Erikson is attempting to isolate.[19] The meaning of "ego identity" and "identity formation"

[17] M. H. Dignan, "Ego Identity and Maternal Identifications," *Journal of Personality and Social Psychology*, **1**, 476, 1965.

[18] Such a review here must be limited and therefore concentrates on those works most pertinent to the issues being pursued here. For instance, there have been many recent discussions of identity formation, some of which were mentioned above. Those of particular theoretical or methodological importance are surveyed here. A more complete listing of identity studies is found in the bibliography.

[19] Some of the many insights and propositions of these writings have already been

is thus greatly expanded by example and usage, resulting in much intuitive understanding of the concept. It is likely that Erikson's future contributions to the study of identity will also be along the two lines of clinical and historical examples, continuing the enlarging file of "case studies."[20] The necessary work of further conceptual and methodological analysis will, one hopes, be taken up by other students possibly stimulated by the wealth of clinical materials now being offered by Erikson and his associates.[21]

Predominantly theoretical identity studies, with clinical illustrations, include those of Blos (1962), Greenacre (1958), Jacobson (1964), Wheelis (1958), Lynd (1960), and Strauss (1959). The first three authors stress psychoanalytic implications of ego identity and often vigorously attack the concept for its many ambiguities. This is particularly true of Jacobson, who in *The Self and the Object World*, points to what she considers a confusion of subjective and objective identity in Erikson's discussions. The concept of the "self" continues to be the major focus of her monograph, identity being treated as it relates to this elusive, perhaps even more complex and ambiguous, concept. Greenacre and Blos, however, are directly interested in the concept of identity. Greenacre is primarily concerned with the subjective "sense of identity," stressing the dual aspects of uniqueness, differentiation from all others, and continuity: "[identity means] . . . an individual or object whose component parts are sufficiently well integrated in the organization of the whole that the effect is a general oneness, a unit."[22]

Less psychoanalytically oriented is Helen Lynd's book, which very sensitively covers many cultural and societal facets of identity formation, drawing upon a broad scope of literary and social science sources. A major portion of the work is devoted to analysis of the relationships of identity formation to issues of shame and guilt. Her notion of identity is wholly derived from Erikson's descriptions and then enriched by her subtle discussions. To be sure, this does not represent a theoretical or method-oriented study. It is a scholarly perceptive exploration of cultural and social aspects of identity.

Strauss's study is an example of a social psychologist's approach to identity. Its perspective is that of social interaction, examining ways in

noted. More extended discussions of these and other assumptions and implied hypotheses are taken up in subsequent chapters.

[20] For example, his *Young Man Luther*, 1958, papers included in *Childhood and Society*, 1950, and most recently, *Ghandi's Truth*, 1969.

[21] See Erikson (1962), Keniston (1965), Lifton (1961), Prelinger (1958), Blaine (1961), Strauss (1963), Wedge (1958), Wheelis (1958), and White (1964).

[22] P. Greenacre, "Early Physical Determinants in the Development of the Sense of Identity," *Journal of the American Psychoanalytic Association*, **6**, 612, 1958.

which personal identity, defined primarily in terms of historical continuity, is supported and demolished. Strauss presents lucid theoretical discussions sprinkled with many illuminating clinical illustrations. His emphasis on continuity as a key variable is particularly meaningful. The idea is implied in Erikson's identity descriptions. However, this facet of identity is not as clearly elaborated by Erikson. Similar to Strauss's presentation, but much less systematic, is Wheelis's (1958) treatment of identity. In addition to observations of sociological and economic changes with their influence on identity development, Wheelis gives detailed and fascinating autobiographical segments as documentation of his theoretical sections. It is through these segments that the "clinical definition" of identity is continued.

While clinical discussions of identity are relatively plentiful, examples of empirical research in identity (other than single case studies) are rarely encountered. Three interesting uses of paper-and-pencil tests for identity measurement are given by Scott and Keniston (1959), Howard (1960), and Dignan (1965).[23] All three instruments employed items initially selected from Erikson's descriptions. Subsequent selections of the test items were made through pretesting and/or panels of clinicians. Covered by these tests were areas such as time perspective, ideology, role experimentation, and sexual identity. Individuals answering the questionnaires were from college groups (Keniston and Dignan) and lower-middle-class teen age girls (Howard). Clinical correlations were attempted in all of the studies. In the Keniston and Scott study, multiple correlations with other paper-and-pencil tests as well as projective tests were determined for each subject. The results, while of uncertain significance, all suggest conformity to general theoretical identity considerations. Thus, for example, Howard found "low identity scores" for "disturbed" girls. Keniston reports high positive correlations between identity scores, other identity scales, and clinical ratings of identity. In addition, Keniston notes negative correlations between identity scores and alienation measures.

In the Keniston study, which was part of a larger project, some of the subjects were followed in subsequent years. Subjects in the other studies were not seen again after the single test administration. Hence, in addition to the caveat regarding the usefulness of this kind of instrument for identity measurement,[24] there is the further problem of whether or not processes such as those of identity formation can be measured through tests administered at a single sitting. Can "an evolving configuration" be accurately detected through one or more determinations taken at the same point in

---

[23] A most recent identity measurement study is that of Martin (1969). The problem of "single sitting," to be discussed below, is also raised by this study's methodology.

[24] That is, that they reveal the most conscious feelings and attitudes of the subjects, hence avoiding important ego identity dimensions, namely, the unconscious aspects.

time? More metaphorically, can a single still picture rather than a movie of someone in motion still accurately capture all the essential parameters of the movement? Clearly, this is an open question, and until there is more systematic identity research of both the "single shot" and longitudinal type it will stay unsettled. The position of this investigator is that the longitudinal approach *is* required for the study of identity formation.

Marcia (1966, 1967) has investigated aspects of identity formation using a combination of interview and paper-and-pencil test techniques. Through applying these instruments to samples of undergraduate men, he describes four "identity statuses" along a series of test dimensions. Several features of his studies differentiate them from the other empirical works. To begin with, Marcia relies on both an interview appraisal as well as psychological tests. Moreover, the investigations focus most specifically on different *types* of identity development. And, finally, the studies include use of an experimental manipulation (of "self-esteem") as one means of differentiating and describing the various "identity statuses." Despite their lack of a longitudinal design, these studies provide a most valuable addition to an obviously limited literature.

They suggest multiple psychological functions which are apparently related to different types of identity formation. Thus Marcia stimulates attention to the careful empirical determination of identity formation and raises questions as to underlying dimensions that might differentiate various forms of identity development. Dimensions of potentially much interest along these lines include cognitive styles, interpersonal patterns, and intra-psychic issues such as self-esteem.

There are at least two other empirical studies pertinent to identity formation. Ross (1962) used specially designed TAT pictures to diagnose and analyze identity conflicts among Indonesians in an acuturation situation. Working jointly with an anthropologist, he compared his partially "blind" analyses of the projective test with life history and other field data gathered by the latter. Favorably high correlations were obtained between the test analyses and field data vis-à-vis identity conflicts. The monograph is of particular interest here in the use made of cultural data to determine various dimensions and etiologies of the identity issues. A second investigation currently in progress (Prelinger and Zimet, 1964) employs, among other instruments, scales derived from ego psychology considerations. The scales, which include intrapsychic and psychosocial variables, are applied to interview and test materials of college students seen at yearly intervals throughout their college careers.[25] Although again it raises the problem of

---

[25] Extensive rationale for and description of these scales can be found in Prelinger and Zimet (1963). A brief summary of the study is in Prelinger et al. (1960).

the sample bias, this approach is most compatible with the one being followed here. Indeed, it meets the "requirements" for identity formation research implied in this short review and critique of the current studies.

The social class research relevant to this study can be divided into three general types on the basis of whether a predominantly sociological-anthropological, purely social structural, or psychological perspective is taken. An outstanding example of the first type is Whyte's *Street Corner Society* (1955). Of more current interest, though with similar orientation, are the studies of Walter Miller (1958). As in Whyte's work, description of lower-class culture is generalized through the observations of small groups in natural settings. Now there are more and larger groups, however, and the number of observers has increased as well. The position that there exist insulated "subcultures" within the United States is implied by Whyte and unequivocally adopted by Miller. A similar assumption of strong coherence and imperviousness to outside influence appears in the studies of "teen culture," age-specific patterns of culture said to cut largely across class lines.[26] The existence of such self-sustaining patterns among class or age groups is by no means accepted by all authors. It is especially controversial with regard to "adolescent cultures."[27] At issue is the degree to which these various subcultures are influenced by other such groupings and by any values and assumptions possibly held by all groups within the society. There is little question by now of differing cultural patterns within a larger society, patterns which are in part related to socioeconomic distinction.[28]

A most recent anthropological contribution to studies of this type is Elliot Liebow's *Tally's Corner*. For the better part of a year, Liebow spent most days, and many nights, with a small group of financially impoverished Negro men. The lives of this street corner group are delineated in *Tally's Corner*. Rather than concentrate on "case" or biographical studies, the book focuses upon those themes and issues common to "Tally" and his friends. Issues such as "being a man" and "being a father without children" are clarified and explored in much rich detail. Interwoven with these issues is a continuing emphasis on relationships within the group and external to it: the men with one another; the men with their "women"; the men with Liebow, a white Jew; and finally the men with the surrounding social institutions, ranging from the "carry-out" shop to the employment agencies with their "bosses." Repeatedly, as Liebow so carefully portrays the culture

[26] Bernard (1961), Coleman (1961), and Havighurst and Taba (1949).
[27] Elkins and Westley (1955).
[28] See the discussions and studies of Stein and Cloward (1958), Barber (1961), Fried and Lindeman (1961), Whyte (1955), Hollingshead and Redlich (1958), and Myers and Roberts (1959).

of this group, there recur in many forms themes of both race and social class.

Also taking an anthropological perspective is the brief and unusual paper of Schatzman and Strauss (1959). Their illuminating study describes social class variation in cognitive patterns and modes of communication. Although derived from a small sample in a western town, evidence is cited from urban populations which lends support for the generality of their findings. In brief, analysis of interviews and spontaneous discussions with lower-class adults about a crisis revealed the following characteristics when compared with similar materials for adults of other classes.

1. Reliance on concrete "particularistic" images
2. Lack of a consistent space-time framework
3. Inability to take the "other" role; all events described from the "I" standpoint
4. Unclear imagery

Confirmation of these findings depends of course on additional more detailed cognitive studies of these populations.

There are a number of more specific and conventional psychological works on lower-class youth.[29] Epstein, for example, reports differences in memories between lower- and middle-class adolescents. The lower-class subjects' memories were characterized by increased sexual, aggressive, and angry incidents plus a later age of recall (eight years old as compared with three for middle class). Consideration is given in the paper to the influence of the middle-class interviewer on both the content and stated age of recall. Korbin also takes up socioeconomic class variables as he reviews the differences in adolescent development between middle- and lower-class youth. Three "common problems of adolescent development" are considered in both groups of subjects:

The first of these problems, "adult authority," is resolved temporarily in lower-class youth by the absence of adults, thus leading to older peer or sibling supervision, if there is any at all. The solution is only transient, Korbin predicts, for the authority problem readily returns—in unspecified forms—in late adolescence and adulthood. The second problem is the dependence-independence ambivalence. With the shortage of nonparental adults, lower-class youths thus have few adults for models. Consequently, they reflect the early independence of the neighborhood, retaining major unfulfilled dependency needs throughout life. Moreover, the absence of accessible adult models leads to the "ideology of the streets" being transmitted relatively intact, rather than that of the family and its specific

[29] See the studies of Epstein (1963), Korbin (1961), and Baittle (1961).

cultural orientations. Finally, there is the phase relationship of dependency and autonomy. Lower-class youth become autonomous sooner than their middle-class counterparts, in that there is less invested in the former as mobility or "security" objects. Whereas independence strivings are thus accepted more easily, however, there again is the problem of managing unfulfilled dependency wishes in the lower-class adult. The outcome of these resolutions for lower-class men is a "primacy of the peer group." Their apparent adult autonomy is marred by a quickened attention given to random types of group support, a tendency that is a direct result of their "unsatisfied psychological hunger for dependency relations." Because of its clear relevance to this study, Korbin's thoughtful paper has been šummarized in some detail. His review is unique in that it is among the few surveys or studies found that specifically examine relationships between socioeconomic class and adolescent psychological development.[30]

The more general area of socioeconomic class and psychopathology or "mental health," has been the object of an increasing amount of popular and scientific interest.[31] As yet it is unclear as to how these primarily epidemiological studies will clarify the relationships and variables that account for correlations between social class and personality development. Using the model of medical epidemiology, we should expect the above works to be most significant in suggesting specific hypotheses concerning mechanisms underlying the statistical trends that are discovered. Further insight then would be dependent upon the quality of the ensuing clinical and/or experimental studies.

Works on the American Negro, our final subject of review, are voluminous. Even when we impose the restriction of direct relevance to questions of personality and culture, their number is still overwhelming. Inasmuch

[30] However, one can find many papers and books about one special group of lower-class youth, the delinquents. There are abundant studies of their psychopathology, psychotherapy, immediate family context, and social dynamics. Yet, even here, when compared with the attention given to the psychology and sociology of delinquency, there is far less analysis of the relationships between individual delinquent development and sociocultural setting. A striking exception to this is I. Chein's *Road to H*, 1964, in which relationships between adolescent deviant development and sociocultural matrix are studied in detail. Another exception is the report of the 1955 White House Conference on Delinquency (Witmer, 1956), in which Erikson, Merton, Redl, and several other sociologists, psychologists, and psychiatrists discuss current knowledge and future research into juvenile delinquency with the concepts and techniques of multiple disciplines. There is also a vast literature on "deviance" in which both sociology and anthropology meet in analyzing delinquents, among other "deviants." Goffman (1962), in a recent addition to this field, uses the concept of identity in discussing issues of deviance.

[31] See Leighton (1959), Srole (1962), Hollingshead and Redlich (1958), Myers and Roberts (1959), Fried and Lindeman (1961), Barber (1961), and Kohn (1963).

as possible, the criteria of selection for discussion here are quality and overall relevance.

Although there were earlier books,[32] Myrdal's *American Dilemma* (1944) stands as a major early work on the American Negro and represents the point of departure for most contemporary studies. Dealing largely with historical and institutional facts of the Negro, the volumes nonetheless do touch on Negro class structure and on some personality issues. There is recognition of the tensions inherent in the matriarchal family patterns, the degraded self-images, and the undischarged aggressions. These subjects are presented by way of anecdotal evidence and general observation, rather than as conclusions of systematic research. In addition to such astute psychological perceptions, Myrdal offers impressively sophisticated insights into the social psychology of American Negro–white relations:

"The 'American Dilemma' . . . is the ever-raging conflict between on the one hand, the valuations preserved on the general plane . . . the 'American Creed,' where the American thinks, talks, and acts under the influence of high national and Christian precepts, and, on the other hand, the valuations on specific planes of individual group living, where personal and local interests; economic social and sexual jealousies; considerations of community prestige and conformity; group prejudice against particular persons or types of people; and all sorts of miscellaneous wants, impulses, and habits dominate his outlook."[33]

Since Myrdal, there have been several important historical and sociological studies of the black American.[34] Of the recent authors, Elkins is of much interest because of his creative use of psychoanalytic and social psychology constructs to reinterpret the system of American Negro slavery. He sees this system as supporting and/or creating a unique American character type, "Sambo": the obedient, pliant, childlike Negro. Especially stimulating is Elkins' choice of the Nazi concentration camp as a model for understanding the closed society of American slavery, vis-à-vis total human control. Raised, but left unanswered, is the question of whether this character type has persisted. Has it remained only as a caricature? Or are there still blacks corresponding to the Sambo type, either in pure form or as variants of it.[35]

Clark (1965), and Hearn (1962) on a smaller scale, discuss the black

---

[32] For example, Phillips (1918) and Dubos (1903).

[33] Myrdal, G., *An American Dilemma*, 1946, pp. xlvii, 1.

[34] See Drake and Cayton (1945), Davie (1949), Frazier (1939, 1957), Elkins (1959), Parsons (1965), Moynihan (1965), and Clark (1965).

[35] We return to this important question in the final chapter of this monograph.

ghetto. Drawing upon his work in preparing the Harlem Youth Opportunities (HARYOU) report and the Negro social psychology literature, Clark presents successive views of sociological, psychological, historic, economic, and political aspects of Negro ghettoes, in particular, Harlem. Parsons (1965) considers ghetto problems in light of a comparison with the assimilation and inclusion of other minority groups in the United States. His analysis is most suggestive. It differs from other discussions of this problem in its argument for the significance of religion as a critical factor distinguishing the Negro from all other minorities. The Negro's historically predominant fundamentalist orientation, his total dedication to otherworldly concerns, was of great significance in implying to society his "incapacity for full participation." Color was but a symbol for the projection of the social anxiety thereby produced by such an orientation. Although this analysis appears incomplete, and at best controversial, it serves as a strong reminder of the unsolved historical problem: Why has the Negro been a minority group so utterly different from any other American minority? This is also essentially the problem that Elkins pursues. Clearly, social science research into contemporary populations will not produce the solution to this great riddle. However, as both Elkins and Parsons demonstrate, the astute application of contemporary psychological and sociological insights may be of unanticipated aid in the reinterpretation of these complex historical processes.

To the historical and sociological dimensions of the black American, Isaacs (1962) adds the parameters of international implications. This is one of the few serious works that confront the question of black identity.[36] Although it is "group identity" that most interests Isaacs, he repeatedly analyzes issues of individual *and* group Negro self-images and their vicissitudes under the impact of African and American social changes. The "elements" of Negro identity are defined to be "name, color, nationality, and origin." Each of the elements are explored in long, thoughtful interviews with black literary and political figures and black participants in "Cross-Roads Africa."[37] Through sensitive and informed analysis Isaacs displays the reciprocal influences present between American blacks and the rest of the world. Toward the end of his discussion he raises the intriguing problem of future development: What will happen as Negroes

---

[36] Erikson, in his recent paper, "The Concept of Identity in Race Relations" (Erikson, 1968), discusses aspects of American Negro experience and its changing dimensions. The article is a series of observations as opposed to any attempt to discuss black identity formation systematically.

[37] A program that sponsored American Negro and white youths for work in and visits to selected African nations during the summer.

"shed the burdens of nobodieness" and "take on the demands and burdens of somebodieness"?[38] What will be the form of future American Negro group identity and their choice of relationship to United States society?

Psychological and social psychological studies of American Negroes are reviewed in several publications by Pettigrew (1964a, 1964b, 1964c, 1965), and Dreger and Miller (1960, 1968). A broad landscape is covered by the many papers in these disciplines, as they range from doll play to IQ-score comparisons, to studies of "self-assimilators." Indeed, one of the serious problems in this area is the very scattering of research efforts frequently leading to either duplication of work or, more often, unsystematic and nontheoretically grounded research.[39] In several discussions Pettigrew proposes outlines of the theoretical framework he thinks will be required to account adequately for the social and individual psychology of American Negroes. He couches his proposed framework in the language of role theory, which on the whole does not stress historical explanation. The papers emphasize the enormous variation in American Negro behavior. He suggests generic variables and situational variables which, taken together, are said to account for the complexity. Although sorely needed, it does not seem that a "psychological theory" of the American Negro is yet available. However, is a new framework or specific social psychology role theory required? It is possible that a framework as complex and general as psychoanalytic ego psychology will be able to deal with data of American Negro behavior and development. Such a thesis is at least worthy of consideration before building new conceptual mazes. Exploration of this possibility is the course taken in the research of Negro adolescence presented in this monograph.

As will be seen, there have been several comparative studies of American black and white populations. Supposed black-white differences in personality development, character, abilities, and psychopathology have fascinated and puzzled investigators. Black and white populations are again contrasted here. But this time the comparison is along very specific lines which have not been used in the prior comparative studies. To be analyzed is an interface between psychoanalytic considerations of individual development and the sociocultural matrix. The interface chosen is an aspect of adolescence in which the individual must synthesize past identifications within himself while he also in some way finds a significant place for himself within the structure of his community. It is this aspect of adolescence in which *consolidation* and *continuity* are crucial issues of maturation.

---

[38] This is similar to the problem raised by Pierce (1968) in his discussion of the increasing demands upon the black American that will emerge upon the diminishing of the deprivations.

[39] T. Pettigrew, *op. cit.*

Some psychoanalytically oriented studies of current desegregation efforts have been reported by Coles (1963, 1965a, 1965b, 1965c, 1967). His studies consist largely of impressionistic observations and informal, sometimes intensive, interviews with both white and Negro children and their families. Most welcome is his inclusion of whites, who become interesting in themselves as well as a needed "control" group.

When we more narrowly seek pertinent literature as being *only* that dealing with black psychological development, in particular adolescence, we suddenly find a sharp decline in available research. The first study specifically of black adolescents was made in 1938 by Dollard and Davis.[40] Two hundred and seventy-seven upper-, middle-, and lower-class New Orleans teenagers were interviewed and 76 chosen as "intensive cases." The latter and their families were given multiple interviews and psychological tests. The results of the study at that time were interpreted as primarily consequences of the differing social classes, caste distinctions being assumed to influence all subjects in an essentially uniform manner.

The same subjects were restudied 20 years later by Rohrer and his associates (1960). In the restudy both social class and self-hate as single factor explanations of the psychological development of the subjects were rejected. Instead, the subjects were found to fall into five major groups of "primary role identifications . . . patterns of cultural identifications, instituted in family life and in the manner of training children." Viewing their subjects primarily in terms of orientation to "one cultural nucleus," be it "matriarchy," "gang," "family," "middle class," or "marginal," is defended by the authors as demanded by the great variations they found when comparing individual developments. The authors readily admit, however, that the analysis is incomplete as it stands.

We are left with a clinically rich study of the social and psychological changes manifest by a selected, heterogeneous group of New Orleans Negroes. Conclusions drawn are modest and cautious. Several of Rohrer's explanatory concepts are similar to the ones of this study. For together with "identification" with cultural orientations, the term "identity" is adopted in order to distinguish the intrapsychic from the cultural dimensions of subjects' experiences. Identity in Rohrer's discussion denotes: ". . . certain comprehensive gains that the individual at the end of adolescence must have derived from his pre-adult experiences in order to be ready for

---

[40] See Dollard and Davis, *Children of Bondage*, 1941. A more recent study is that by Pierce (1968). It is primarily a series of speculations and concerns about the effect of the now lessening deprivations and degradations. As the black "underprivileged" now become privileged, what will the impact be on black adolescents? It is recalled that Harold Isaacs raises a similar more general form of this question.

the task of adulthood."[41] This term, however, is a minor one in the subsequent presentations of results, and when used it is taken as roughly synonymous with self-image. The limited generalizations and mixture of social psychology, anthropology, and psychoanalytic concepts is in part a consequence of the study's "team" approach of anthropologists, psychologists, sociologists, and psychoanalysts. Such a group obviously was productive of a multifold perspective on the data. Yet one suspects that it also hindered the use of any single theory as well as the formulation of hypotheses from the data. There are many points of convergence between Rohrer's work and this study of New Haven adolescents. The most important difference lies in our adherence to a single theoretical framework for the guiding and interpretation of the investigation.

Kardiner and Ovessey (1952) also intensively studied black adults and adolescents. Their work differs from Rohrer in both method and interpretation. Using modified psychoanalytic interviews, they examined 25 New York Negroes of varying age, sex, and class. On the basis of their detailed case histories, they conclude that there are conscious and unconscious trends of self-hatred and "identification with the white" in all of their subjects. In evaluating their conclusions it is important to consider the fact that half of their subjects were also explicitly "patients," receiving free psychotherapy instead of cash for being informants. Moreover, Rohrer, in reviewing Kardiner's case records, claims that there is evidence of conscious self-hatred in only seven subjects, five of whom were patients.[42] The existence of self-hatred in all American Negroes obviously cannot be convincingly argued from numerically limited and then probably psychopathologically biased samples. Only through observations of many black subjects from various health, geographic, and class backgrounds can we answer this question, or the more general one of generic responses and tendencies among American Negroes.

Studies of black developmental problems and adult psychopathology are of course valuable in suggesting specific conflicts and trends that may characterize certain groups of, or possibly all, American Negroes.[43] There are some findings common to all of these studies: the black subject or patient is described as having low self-esteem, marked unconscious aggression and hostility, and a degraded self-image. Sources of data range from free association interviews to the Thompson Picture Arrangement Test (PAT). Karon's study is distinctive in using the latter "standardized" projective instrument, and in his emphasis on the differences between northern

[41] Rohrer and Edmondson (1960), p. 86.
[42] Rohrer and Edmondson (1960), p. 72.
[43] Rose (1956), Dai (1955), Powdermaker (1955), Sclare (1953), and Karon (1958).

and southern Negroes. The latter differentiated themselves from their northern counterpart on the basis of greater concern with aggression, weak and labile affect, and conflicting work motivation.[44]

Probably the most difficult productions to study systematically are the expressions of Negro literary figures. These writings are nonetheless of inestimable importance for insights that may not be gained through other means. Popular writers such as Silberman (1963) have surveyed many of the literary productions. White (1947) performed a value analysis of *Black Boy*. Bone (1958) studied the works of several Negro writers. In general, however, these literary works have not been seriously analyzed. The essays, books, and plays are a rich and largely unmined source of important data and understandings about the American Negro.[45]

A literary form of special interest to the study of identity formation is the autobiography. At its best, the autobiography provides the chance to trace individual trends of development, their internal consistencies, their thema, and the articulation of these trends with social and cultural patterns. A number of Negro autobiographical works have appeared in recent years.[46]

Clearly, these autobiographical works represent a growing "file" of case studies whose careful analysis can only contribute needed data about Negro personality development. In *Manchild in the Promised Land*, Brown richly describes a childhood and adolescence spent in a ghetto, followed by prison. He notes a series of changes in his life style and orientation, touching occasionally upon possible determinants of the shifts. It is likely that further analysis of the material would lead to other formulations about his development. Such a possibility exists for each of the autobiographies cited.

This comparative study of black identity formation has been informed

---

[44] *Black Rage* is a more recent study of American Negroes with a dominant clinical focus. Through much rich interview material and more general social commentary several of the issues found in the above writings and in *Tally's Corner*, are elaborated upon. See Grier and Cobb (1968).

[45] Best known of the current articulate Negroes is James Baldwin. His prolific works are filled with themes of protest, injustice, emasculation, and degradation. These themes are developed in all the novels as is, to a more variable extent, the explicit problem of being Negro; see *Go Tell It on the Mountain, Another Country*, as well as the collected essays in *Nobody Knows My Name* and, more recently, *The Fire Next Time*. MacInnes (1963) is an interesting summary and discussion of Baldwin's writings. Other important contemporary Negro writers and writers on Negroes include Wright (1945), Redding (1951), Ellison (1952), Miller (1959), Hansberry (1959), Jones (1967), and Fanon (1962).

[46] These include Malcolm X, *Autobiography of Malcolm X*, 1967; J. Baldwin, *The Fire Next Time*, 1963; E. Cleaver, *Soul on Ice*, 1968; S. Redding, *On Being Negro in America*, 1951; and C. Brown, *Manchild in the Promised Land*, 1965.

in many ways by the essays, novels, and autobiographies as well as by the more "objective" empirical research. There is a large and ever-expanding literature on black Americans. From this literature a series of significant contributions were selected and are discussed in this chapter. Generally omitted from more detailed review were the fictional and autobiographical works. Such an omission is, however, in no way related to their importance. It does testify to the difficulty of treating these productions. One approach could be analyses of each work, the results of which might then generate new hypotheses and formulations. Alternatively, systems of content analysis might be directly applied to each specific work as means of investigating particular hypotheses and issues.[47]

Within the group of empirical studies, there are three that can be said to have immediate bearing upon this research: *Children of Bondage, The Eighth Generation*, and *The Mark of Oppression*. It is only in these efforts that black adolescents are a focus of study as well as discussion. Although the present study differs from these in terms of both theoretical bases and techniques, they nonetheless provide valuable comparative data.

At this point, then, we have an overview of the theoretical and empirical background germane to the study of black identity formation. With this perspective in mind, we can proceed to full examination of the patterns of identity formation that were studied in the New Haven adolescents. The remaining chapters deal with all aspects of this investigation, moving from the study itself to speculative models for understanding the discovered patterns of identity formation.

---

[47] Such a system for studying "heroes" and "antiheroes" is currently being developed.

CHAPTER 2

The Structure of the Study

THE SAMPLE AND TIMING OF THE INVESTIGATION

The names of possible subjects were obtained from New Haven junior and senior high school guidance counselors and assistant principals. These school officials were requested to choose students who had the following characteristics: male, entering sophomore year of high school, from a lower socioeconomic class family,[1] and neither delinquent, predelinquent, nor "college-bound." The boys were then seen individually and accepted for the study on the basis of the above criteria and their motivation to participate. A total of 23 Negro and white boys were initially interviewed and tested. From this group five were not seen again after the first year; other "dropouts" occurred in later years. Table 1 summarizes characteristics of all subjects who participated in the study.

In brief, the sample consisted of 11 Negro, 11 white, and 1 Puerto Rican boy. Their chronological ages at the start of the project varied between 14 and 16 years old; there was no significant difference between the average age for either major group. Socioeconomically, 22 of the 23 boys were from Class 4 or 5 families, as determined by the Hollingshead (1957) social class index.[2] Eight of the 11 Negroes were Class 5 ("lower-lower class"), while 4 of the 11 whites were from this class.[3] Of interest in regard to attrition

[1] All boys who on the basis of the Hollingshead (1957) index were from a Class-4 or -5 family, were considered to be "lower socioeconomic class."

[2] This index is intended as an objective, easily applicable instrument for determining the positions individuals or households occupy in the status structure of our society. It is based on the head of the household's occupation and amount of formal education. Rationale for the index and further description of its application can be found in Hollingshead (1957).

[3] It is possible that the Hollingshead index is not applicable to nonwhite groups without some modification in either scaling or factor weighing. Brown (1955) and Glenn (1963), among others, have suggested that the symbolic prestige value for both education and occupation may be very different among Negroes when compared with white populations. Roughly, the same job or education is seen by Negroes as "worth

Table 1. The Sample

Subject	Age[a]	Race[b]	Socio-economic Class[c]	Number of Years in Study[d]	Head of Family
FM	15.0	N	4	3	Father
GR	14.5	N	5	3	Father
LT	15.0	N	5	3	Father
OS	14.0	N	5	1	Father
FL	14.0	N	5	2	Mother
MD	16.0	N	5	3	Mother
TM	15.0	N	5	3	Mother
BA	16.0	N	5	3	Mother
KE	14.5	N	5	3	Father
JE	15.0	N	4	2+	Father
KB	16.0	N	4	1	Father
HN	15.0	PR	5	3	Father
AL	15.0	W	4	2	Father
NS	15.0	W	5	3	Father
BT	14.0	W	4	3	Father
ND	16.0	W	5	1+	Father
VR	15.0	W	4	3	Mother
JK	16.0	W	4	3	Father
JR	15.0	W	5	3	Father
IQ	16.0	W	4	1	Father
CS	15.0	W	3	1[e]	Father
LE	15.0	W	4	1	Father
DM	15.0	W	5	1	Father

[a] Age at start of study, to closest half-year.
[b] N, Negro; PR, Puerto Rican; W, white.
[c] According to Hollingshead index; see footnote 2.
[d] Three years represents completion of study.
[e] Dropped from the study because of both his social class and college goals.

is that all Negro dropouts (from the study) and 3 out of the 5 white dropouts were from Class 4 families ("upper-lower"). This is but one of several interesting features that characterized the subjects who left the study. These subjects are described in greater detail in Appendix B.

Geographically, the subjects came from many areas of the city; from Rockview and Elm Haven Project, Dixwell Avenue, Wooster Square, Fair Haven, and central New Haven. Explicit effort was expended to avoid choosing the sample only from "ghetto" areas. Equal numbers of subjects

more" than it would be in a similar group of whites. To date, no systematic modification based on ethnic or racial differences has been derived for the Hollingshead index, or for any other social class index. Obviously, this is sorely needed. For the data here, it is likely that we have a "false low" for the Negro families. Probably then, the Negro and white boys are from approximately the same socioeconomic class.

came from each of the two public high schools; one subject attended a state trade school. Other features of the sample, such as family composition, geographic mobility, and school performance are presented in Appendix B.

The sample, then, is a highly specific one. It is, to begin with, all male. In addition, it is limited to members of lower socioeconomic classes. These constraints were self-imposed for very specific reasons. One purpose of the study was to investigate *non*-middle-class youth in terms of identity formation, hence the limitation to lower-class populations. It would be of much interest to make comparisons across class lines using the techniques of this study, as well as across ethnic ones. Such considerations are further discussed in Chapter 7.

A second constraint of the sample is that of sex. There is no reason to assume that identity formation is similar for males and females. However, rather than study possible intersexual differences in these developmental processes in addition to possible interracial ones, the choice was made to limit the sample to males and thereby study one specific contrast within an otherwise generally homogeneous group. In other words, sex and social class were held constant in the group of subjects, while race was selected as the contrast characteristic.

All subjects were paid at the rate of one and one-half dollars per hour. Initially, and then twice each year until graduation from high school, each boy was seen for a combination of interviews and tests.[4] The schedule of interviews and tests of this first year and all subsequent ones is given in Table 2.

Generally, interviewing and testing were carried out individually at neighborhood schools. When this arrangement was not feasible, an interviewing room with built-in tape-recording equipment at Yale Medical School was used. Whether seen in the latter facility or at the schools, all subjects were informed that the interviews were being recorded and that the recordings would be kept anonymous.

THE TECHNIQUES OF INVESTIGATION

The Interviews

All interviews were tape-recorded and later transcribed. Inasmuch as possible these sessions were unstructured, the interviews being guided only by a list of topics to be covered. These topics were largely explored through

---

[4] These tests included the TAT, Sentence Completion, Draw-a-Person, Prelinger-Otnow Intimacy Test, and *Q*-sort. All tests, except for the *Q*-sort and Draw-a-Person, were given at the start and at the completion of the study. The *Q*-sort and Draw-a-Person were administered *each year*, in the spring.

Table 2. Schedule of Procedures

Year	Dates	Interviews	Tests[a]
First	August–September	Initial interview Work Heroes and myths Body image and intimacy	Sentence completion Draw-a-Person TAT Intimacy test Q-sort
	June	Interim interview	Draw-a-Person Q-sort
Second	October	Interim interview	
	June	Interim interview	Draw-a-Person Q-sort
Third	October	Interim interview	
	Spring	Interim interview and repeat topics of all initial interviews	Repeat all of initial tests, including Q-sort
Fourth+		Follow-up interviews	Follow-up Draw-a-Person and Q-sort

[a] These tests were given in the order described here. However, it is the results of the Q-sort that are presented in detail in the text. The other results are not reported here. Description of the other procedures can be found in Machover (1949) (Draw-a-Person), Dorris et al. (1954) (sentence completion), and Otnow (1962) (intimacy test).

open-ended questions which were introduced during the initial set of interviews and all later "interim" interviews. Thus while the interviews were "free flowing," inclusion of a standard set of topics in each set of interviews was assured. The interim sessions were devoted to reviewing recent experiences and events as well as new plans, wishes, ideas, conflicts, and decisions that had occurred since the last interview. Historical matters were discussed at length where appropriate during some of these sessions. Usually, the course and content of the interim interview was determined by the subject as he brought the interviewer up to date and chose to pursue related new issues.

The interviews and tests were all conducted by the same interviewer[5] who was first identified to the subjects as a medical student interested in studying "how teenagers change." It soon was apparent that this information did not completely clarify who he was and what his purposes were. Repeatedly, subjects showed confusion and distortions regarding his professional status and worries about his motives.

As might be anticipated, the most extensive clinical data of the study

[5] The interviewer and tester throughout the study was the author.

was generated from these interviews. The material was studied in three ways: by impressionistic analysis for prominent themes and individual patterns of development; by application of objective scales and codes by independent judges; and, finally, as a means of generating statements for the yearly Q-sorts. The Q-sort and thematic analysis are those procedures most directly related to the results reported in this monograph. It is therefore important that we now examine the nature of this valuable instrument, the Q-sort.

The Q-Sort[6]

From each interview all explicit self-descriptive statements made by the subject were abstracted. These statements concerned his attitudes, plans, wishes, fantasies, feelings, judgments, actions, or thoughts. Each of the statements was then phrased in the simplest possible form, that is, as a positive statement in the present tense and without qualifiers or conditional clauses. The statement was then placed on an individual three-by-five card. The complete pack of such cards became the subject's "deck."[7]

Each year, within one to two weeks of the set of interviews, the subject was presented with his deck of statements. He was told that this numerically coded set of cards was based on what we had discussed during our several conversations. Ten four-by-six cards numbered individually with large numbers from 0 to 9 were laid out on the table in front of him, in ascending order. He was then requested to judge each statement in terms of how well it described him, placing the most accurate descriptions at the 9-position, labeled "most important," and the least accurate ones at the 0-position, labeled "least important." He could place those statements that he felt did not belong at either extreme at any of the intermediate positions (1 to 8), depending on which extreme to which they were closer. Or, statements could be judged as neutral and placed on the number 4- or 5-

[6] The specific use of the Q-sort in this study is an elaboration of its application by Prelinger, as described in his NIMH Grant Proposal M-3642. The help of Dr. Prelinger in developing and using the technique for this study was considerable and is gratefully acknowledged.

The Q-sort itself, and the more general methodology from which it emerges, has been summarized and most thoroughly discussed by Stephenson (1953). This work provides a basic introduction to the method. The orientation of Stephenson, originator of the Q-technique and methodology, is aptly characterized by Brown (1968): "Stephenson has been most concerned from the start with providing an objective approach to the problems of subjectivity" (p. 589). Brown's (1968) comprehensive bibliographic review of this entire area gives clear evidence of the extensive and varied applications that have been made of this technique in the past 30 years.

[7] Examples of these statements are given in Appendix D.

position. Hence the subject was required to sort his statements on a 10-point scale of importance; the only distributional requirements being the use of each of the two extreme categories at least twice.

Using the same deck or its exact duplicate, the subject then sorted the cards seven additional times, once for each of seven other self-images. Always in the same order, the self-image instructions were introduced one at a time for seven more sorts. In summary then, the subject was told to arrange the deck of statements according to how well they described him for the following possibilities:

1. How I am now
2. How I would be if I were a perfect son to my mother
3. How I appear in the eyes of my friends
4. How I will be in ten years
5. How I would be if I were a perfect son to my father
6. How I appear in the eyes of other people
7. How I was at the beginning of junior high school
8. How I would be if everything worked out exactly the way I want it to; how I would be if all my dreams came true

This procedure was first performed following the initial set of interviews, in the fall of sophomore year of high school. Thereafter, it was repeated in June of every year using the subject's original deck of cards, to which had been added all new self-descriptions made in the interval interviews. Thus the deck "grew" annually. No statements were ever deleted. For all subjects who completed the study, a total of four complete sets of sorts were obtained.

Two different product-moment correlations can be derived from these data.[8] First, the correlations of each year's self-images with one another can be calculated. A total of 28 such correlations were determined each year for each subject. These correlations are *intrayear* correlations and represent the degree to which the subject's self-images resemble or differ from one another in any given year; they are estimates of the "fit" of the images.

The second kind of correlation is related to change over time. The change in a self-image was determined by comparing the arrangement of statements it elicited in any two different years. Suppose, for instance, that we

---

[8] These correlations, as well as all average correlations, were computed through the IBM 7040/7049 system. Additional information from these sorts includes the *content* of those statements that are consistently rated at the extremes for individual subjects. The results of such an analysis will be presented at a later date. Both sets of analyses were greatly facilitated by the use of the Yale Computer Center and the able assistance of Roger Bakeman, programmer at the Center.

want to compare the individual's description of his current self (Sort 1) in the first year of the study with the same self-image two years later. To accomplish this comparison quantitatively, the correlation between the sort of his cards for this image in Year 1 and in Year 3 would be calculated. This type of correlation is an *interyear* correlation; it represents the exact quantitative determination of the change or constancy of any given self-image.

Through this *Q*-sort technique and these forms of quantitative analysis, the identity formation of the black and white subjects was studied. Now the relationship between the technique and identity formation must be clarified, and in describing this relationship the operational definition of identity formation is fully elucidated.

CHAPTER 3

The Measurement of Identity Formation

PRELIMINARY CONSIDERATIONS

A definition of identity formation taken from one of Erikson's recent papers was quoted in Chapter 1. The notion of identity formation as an "evolving configuration" of intrapsychic and psychosocial variables appears in various forms throughout Erikson's writings, be they clinical, historical, or theoretical. Generally, these formulations are well suited for intuitive understandings of the case studies and essays and even at times for theoretical clarifications. To develop means of objective measurement to permit the design of replicable studies of identity, however, these descriptive statements must be transformed into a more exacting composite definition which can be used by clinical and nonclinical investigators. The translation of descriptions of identity into a more "operational" statement is discussed and proposed in this chapter.

The configuration or "integration" of components that Erikson calls "identity" is not restricted to adolescence. Development of identity has been progressing through all prior psychosocial stages. In adolescence, however, the process becomes problematic. It becomes increasingly conflictful for a variety of intrapsychic and psychosocial reasons, foremost among them being: the onset of puberty, bringing with it renewed oedipal conflicts and the demands of increased libidinal drives; cognitive changes such as increased awareness of irreversibility; and new societal demands in terms of work, sexual commitment, and ideological commitment. This "primacy" of ego identity is manifested by the individual's conscious and unconscious preoccupations with consistency, personal and social significance, commitment, and irreversibility. Most inclusive of these concerns is that of personal continuity: a sometimes intense sense of physical, ideological, social, and emotional change. More concretely, a heightened awareness of continuity and "wholeness" appears through themes of body image, sexual definition, social roles, values and ideals, and interpersonal intimacy.

"Successful" resolution of the task of identity formation is said to be

29

heralded by the individual's gain in sense of direction, meaning, and overall personal and social coherence. Concomitantly, there is a decline in his preoccupation with synthesis, continuity, and the other themes noted above. One other sign of successful resolution, predicted by Erikson, is the emergence of the capacity for and interest in sustained heterosexual intimacy: the critical psychosocial task of the next maturational stage.

However, continued or intensifying conflict over problems of personal continuity, unremitting preoccupation with identity themes, or the inability to attain interpersonal intimacy are all signs of nonresolution and "identity diffusion." Another, probably one of several, pattern of nonresolution is that of identity foreclosure. This has surface resemblance to successful consolidation. There is a diminution in all of the above conflict areas. But this is where the similarity ends. Rather than synthesis and richer definition, the result is impoverishment and restriction. The conflicts and confusions have been lessened by avoidance. The individual's sense of direction and self is forever fixed, as it is in response to a withdrawal from conflict areas, a rigid closing off of possibilities. Intimacy in this state is neither possible nor desirable.

Following the operational definition of identity formation, these "identity dynamics" are reconsidered in more specific empirically defined terms.

AN OPERATIONAL DEFINITION

Identity formation is a construct that refers to a particular phase in ego development. The nature of this step in ego development is closely dependent upon social and cultural contexts. These factors must be considered in fully analyzing any given case of identity formation. When dealing with the problem of the means for objectively specifying and measuring identity formation in a given individual or group of individuals, however, such contextual matters can and must be set aside. They become relevant following determination of identity formation itself. It is then that the questions of changes, relationships to other ego identity patterns, and deviations from predicted development can be raised. Answers to these and similar problems require that social and cultural processes be taken into account in addition to the psychological ones.

The advantages of formulating a workable index for identification and measurement of identity formation are more than obvious. The host of hypotheses implied in the clinical studies can gradually be tested by investigations on many and varied populations. The validity of the inferences derived from this clinical construct can be evaluated and, in addition, further theoretical implications clarified and discovered.

In this section the operational definition of identity formation is first detailed. After the full statement of the definition itself, each of its components receives further discussion.

Identity formation designates specific processes involving an individual's self-images, namely: (1) their structural integration, and (2) their temporal stability. Changes in either of these processes is indicative of specific types of identity formation.[1]

It is necessary to examine each term of this definition to further clarify its empirical meaning:[2]

Self-Image

The term self-image refers to those concepts, conscious, preconscious, and unconscious, by which an individual characterizes himself. Identity has often been described as referring to "consolidation" or "integration" of multiple components of ego function such as capabilities, significant defenses, identifications, and ideals. The term "self-image" includes these other components. In other words, self-image is a higher level of abstraction; each self-image may be further analyzed into the above elements. To understand a given individual fully, to come to grips with the significance of his identity as part of his overall personality development, it is necessary to investigate the more traditional intrapsychic ego functions, as well as those elements included in id, super ego, and ego ideal constructs. However, the notion of identity formation qua identity formation does not *directly* concern itself with anything but self-images and in particular the two basic aspects as

---

[1] This is a *formal* definition, specifying those general qualities—criteria—that indicate the presence, absence, or current state of the ego development called identity formation. There are *content* aspects also. Certain content variables, psychosocial issues and themes, are postulated as being related specifically to the development of ego identity in adolescence. Such issues include preoccupation with body image, sexual prowess, role definition, values and ideals, and conflicts over interpersonal intimacy. They are "markers," reflecting the psychosocial events associated with identity at this point in the life cycle.

Since ego identity develops throughout the life cycle, it is therefore possible that different issues or "themes" specifically linked to it are prominent at other times. This possibility has never to our knowledge been investigated. Furthermore, not only are the issues "phase specific," but they also may be culturally and socioeconomically conditioned both in terms of quality and intensity. This latter proposition is suggested in following the development of the two boys discussed in Chapter 6. Much of the data there is related to ethnic and adolescent factors in the identity development of two subjects.

[2] Reflections about this, and the subsequent operational definitions, were greatly facilitated through discussions with Dr. Ernst Prelinger; particularly in relation to his use of the Q-sort.

noted.[3] There are probably many techniques for objectively determining the array of a person's self-images. Obviously, psychoanalysis and psychotherapy are means for this. A more immediately objective and empirically verifiable one is by means of the Q-sort. As described earlier, through this technique an individual sorts—in the sense of rating—given groups of statements about himself under specific conditions, such as his self-image now, or his self-image last year, or in 10 years. The statements can be a statement set used for all subjects, or they can be those taken from the subject's various self-descriptions. The latter method probably assures that the elicited sorts most closely correspond with the subject's verbally expressed self-images.

Structural Integration

The many clinical definitions by Erikson and others emphasize the "consolidation" processes of identity formation. There is a coherence of elements, a growing "organic" whole. Translating the valuable intuitive impression to a measurable function gives the concept of "structural integration." By this is meant the intercorrelations of an individual's self-images at any given time as measured by the Q-sort. These are the *intrayear correlations* described in Chapter 2. Structural integration, then, is measured by the average of all these intercorrelations at any specific point in time. When dealing with more than two self-images, as is almost always the case, one can also speak of those more-or-less closely integrated images suggesting a hierarchy of importance. Finally, if the same subject is seen over any period of time, he may be retested for changes in this identity formation process.

Temporal Stability

Development of identity is also reflected by changes and consistencies of self-images. When speaking of identity formation, the concept of *continuity* is frequently used. One important meaning of this is constancy of self-images: the extent to which present self-images resemble those the individual held in the past. In these clinical descriptions an "increasing sense of continuity" is taken as an important indication of identity formation; a decrease in this sense suggests the opposite. Again, this intuitive

[3] In his most recent discussion, Erikson (1968) notes that identity formation can have a "self aspect" and an "ego aspect." He notes that "self-identity" refers to the integration of the individual's self and role images. In these terms then, this is a formal operational definition of the "self aspect" of identity formation.

impression can be operationally defined by the Q-sort. The change or constancy of any specific self-image can be measured by obtaining the correlation between the sorts made under the same conditions at any two different times. Such determinations are *interyear correlations*. This can of course be done with any self-image, providing the conditions of sorting and number of statements are held constant. For any given time interval, the temporal stability process of an individual's identity formation is determined by the average of all interyear correlations.

This process can obviously only be measured over time, once more reinforcing the proposition that identity formation must be investigated by a longitudinal approach. It is conceivable that some observations and indices relevant to identity formation can be obtained by a single study. As more is learned of identity development, these partial determinations should become increasingly useful in assessing states of identity formation. However, such data are currently of questionable significance in the study of identity formation.

VARIATIONS IN IDENTITY FORMATION

Using the definition presented above, the variants of identity formation observed by Erikson can be recast into operational statements. Individuals can then be typed into these clinically useful categories on the basis of clear, publicly agreed upon criteria. There is obviously no reason why one must adhere to these clinical types; the study of other populations using the Q-sort and similar measurements of self-image interaction may suggest even more useful categories or, more probably, finer distinctions within these. Erikson's classifications are employed here, however, primarily because of their prior utility in clinical discussions and the amount of pertinent clinical data that he and others have collected.

Progressive Identity Formation

Throughout childhood and adolescence ego identity develops, with an acceleration of this process occurring in adolescence. Less specific remarks have been directed toward the period after adolescence. It is implied, however, that the process slowly continues, reaching a climax in old age, the "wisdom" stage of the life cycle.[4] We may describe an individual -as

[4] See E. Erikson, "Growth and Crises of the 'Healthy' Personality," *Psychological Issues*, **1**, p. 98, 1959 and Erikson (1968).

manifesting progression in identity formation when *both the structural integration and the temporal stability of his self-images are simultaneously increasing over any given period of time*, though not necessarily at the same rate.[5]

Identity Diffusion[6]

This clinical type designates those cases in which there is failure to achieve the integration and continuity of self-images. The category is a broad one and probably includes several subtypes, for there are conceivably multiple etiologies underlying this outcome of identity formation. Such a state may be present at any stage of the life cycle. However, it is theoretically most likely to be manifest at adolescence. Again, only through further empirical investigation can this hypothesis of stage specificity be studied. For example, does identity diffusion occur during latency or unexpectedly in adulthood and also hinder continued psychosocial development?

Operationally, an individual is said to be in a state of identity diffusion when *both processes of his self-images show a repeated decline in magnitude over any given period of time*. Again, these changes need not be at the same rate. In fact, differing rates may be one sign of variant kinds of identity diffusion. However, of the two processes, structural integration is the most critical here in defining the state of diffusion. Another possibility in this category is a form of *"attenuated diffusion"* in which either (1) structural integration decreases and temporal stability remains constant, or (2) a milder type in which temporal stability decreases and structural integration remains constant. It is possible that these represent early forms of flagrant identity diffusion in which reversibility is more likely.

Identity Foreclosure

This variant in identity formation superficially resembles progressive identity development. There seems to be present a sense of integration, "purpose," stability, and a diminution in subjective confusion concerning these matters. However, the stability and purpose are the result of an avoidance of alternatives, of a certain restrictiveness which eliminates ambiguities. What appears to be the outcome of a successful progression in

[5] Formal symbolic statements describing this development, as well as the variants, are given in Appendix C.

[6] In his most recent discussion, Erikson now refers to this variant as "identity confusion." We retain the previous, *more familiar* term here, however; see Erikson (1968), p. 131.

identity formation is actually an impoverished, limited self-definition and sense of continuity. The strains and confusions inherent in the syntheses necessary for adolescent identity formation have been bypassed, as the developing individual has settled upon a certain identification or set of identifications as forever characterizing himself in all ways.[7] Identity foreclosure is thus an interruption in the process of identity formation. It is a premature *fixing* of one's self-images, thereby interfering with one's development of other potentials and possibilities for self-definition. An individual does not emerge as "all he could be."[8]

Operationally, a person is said to manifest identity foreclosure when *either or both the structural integration and temporal stability processes of his self-images remain stable*, or only temporal stability shows continued increase while structural integration is unchanging.[9]

Negative Identity

In this identity formation variant the configuration of self-images is fixed upon those identifications and roles that have been presented to the individual as most undesirable. The individual sense of identity is based on the repudiated, the scorned. With its emphasis on those historically repudiated and rejected identifications, this variant represents a premature closing off of new syntheses of identifications, of new identity configurations. As such, it is a type of identity foreclosure. Although theoretically negative identity has different developmental roots than identity foreclosure,[10] the structural pattern is the same: abortive identity formation, premature fixation of self-images, thereby halting further evolution of self-definition.

It follows that the operational definition of negative identity is the same as that of identity foreclosure. Structural integration and temporal continuity are unchanging, remaining at the same level year after year. Or if there is any change, it is in the temporal continuity process that increases

[7] This is not to imply that the avoidance of these issues is the determinant of identity foreclosure. It is but one consequence. As can be seen in Chapters 6 and 7, the determinants of this variant are indeed a complex matter.

[8] E. Erikson, "Growth and Crises of the 'Healthy' Personality," *Psychological Issues*, **1**, 1959, p. 87. In that section Erikson gives examples of "good little worker" or "good little helper" as illustrating such premature fixing.

[9] It is conceivable that this structural integration could be at a very low level of correlation and thus be more representative of an interrupted identity diffusion. The definition of foreclosure may therefore require a numerical specification such as: both structural integration and temporal stability must be .40 or greater.

[10] Theoretical aspects of the genesis of both *identity foreclosure* and *negative identity* are discussed in Chapter 7.

its level of activity and thus reflects even greater similarity of self-images over time. To differentiate negative identity as a type of foreclosure, the actual *content* of the self-images must be examined both with respect to one another and, even more importantly, with respect to knowledge of the individual's own history.

Psychosocial Moratorium

An individual is described as being in a psychosocial moratorium when he is "finding himself," experimenting with varied roles, new self-images, and future plans, at all costs remaining uncommitted to any particular alternatives of self-definitions. At the same time he is *not* tending toward, or experiencing, a type of identity diffusion. This is, of course, the *antithesis* of foreclosure. Rather than rigid sameness, the content and patterns of self-images show continual variation. The key concept here is "openness," noncommitment. No irreversible decisions or plans are made. A partly conscious, partly unconscious attempt is made to ensure maximal flexibility and diversity before the further elimination of any possibilities, alternatives, or actions inherent in all stages of identity formation. Operationally, an individual is in the period of *psychosocial moratorium when the temporal stability and structural integration of his self-images shows significant fluctuations (increasing and decreasing) over a given period of time.* Consistent with the tendency running counter to diffusion, the structural integration feature shows less fluctuation than that of temporal stability, particularly in terms of any decrease in value.

Implicit in all of the preceding definitions is the notion that identity formation is a developmental process. The criteria specified by each operational statement require *time-based comparisons* of correlational levels. According to this position determination of any type of identity formation rests upon observations taken at several points in time. It follows that there are no absolute criteria with which one can define identity formation variants. Even more specifically, observations from a single point in time cannot by themselves adequately characterize this *developmental* process. It is a patterning over time of particular variables that serves to define the type of identity formation taking place.[11] In interpreting the data presented in the next chapter we on several occasions need to recall this key notion of time patterning.

[11] It is this insistence on multiple observations over time that most separates this study from the few other empirical studies of identity development. Keniston (1959), Dignan (1965), and Marcia (1966) all rely on single testings for their analyses of identity types.

The operational definitions of identity formation variants and their clinical counterparts, are summarized verbally and graphically in Tables 1 and 2. After this review of the kinds of identity formation, we turn to examine the application of these empirical definitions to the longitudinal data. In Chapter 4 are answers to the question, *What* were the identity formation patterns found for white and Negro boys?

Table 1. Identity Formation Variants

| Variant | Definition | |
	Clinical	Operational
Progressive identity formation	Continual increase in synthesis of identifications, personal and social continuity. "An evolving configuration of constitutional givens, idiosyncratic libidinal needs . . . consistent roles"	Consistent increase in both structural integration and temporal continuity
Identity diffusion	Continual decline in synthesis of identifications and ego functions; decreasing sense of wholeness and continuity with self and community. Fragmentation	Progressive decline in structural integration and temporal stability; more severe form has greater decline of structural integration
Identity foreclosure	Rigidity in self-definition. Lack of any change in synthesis of identifications or other synthetic processes. Characterized by premature aborting of identity development. At first glance resembles "successful" identity formation	Little or no change in temporal stability or structural integration. If there is any marked change, it is in the direction of increasing temporal stability, in face of unchanging structural integration
Negative identity	Premature self-definition based on *repudiated,* scorned identifications and roles. Commitment to what is personally despised	Same as identity foreclosure, of which it is a subtype. Examination of *content* of self-images essential to distinguish it as a subtype of foreclosure
Psychosocial moratorium	"Experimental" state; no firm commitments made. "Trying on" of roles and integrations characterized by flexibility, flux, but *not* disintegration	Fluctuation, swings, in both directions of temporal stability and structural integration

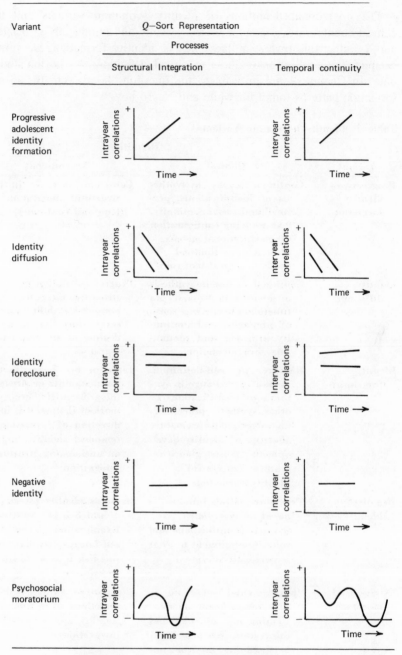

Table 2. Identity Formation Variants.

CHAPTER 4

Black and White Variations
in Identity Formation

INTRODUCTION

A group of black and white adolescents were individually followed from the start to the completion of high school.[1] In this chapter the results of the quantitative Q-sort studies are presented and subsequently analyzed in terms of identity formation patterns.

Four sets of self-images are discussed: parental self-images, personal time self-images, fantasy self-images, and current self-images. Relations within and between these sets are reviewed by examining annual interyear and intrayear correlations and changes in these values over time. Following the survey of these self-images, we consider the specific indices of identity formation: the average interyear and intrayear correlations.

In this chapter we focus on the presentation of analyses and findings; the three chapters that follow deal at length with interpretation.

By design, the chapter is a summary of numerous correlations from each of the subjects. The subjects sorted large decks of cards annually for four years. On these cards were specific self-descriptions of many kinds,[2] which each subject had given in the prior interviews. The cards were arranged— sorted—by the subject in terms of how accurately they described each of eight self-images. Hence every year a subject resorted his set of cards to describe these eight self-images.[3]

---

[1] This group consisted of 22 subjects at the outset of the study. With varying attrition rates, the number varied over the subsequent years. In the significance tables included in Appendix A, the exact number and racial composition of subjects per year interval are given.

[2] Chapter 2 gives general descriptions of these statements. Illustrations of specific statements and how they were arranged are presented in Appendix D.

[3] The specific self-images are detailed in Chapter 2 and also in later sections of this chapter. The way in which this Q-sort was used here, and most specifically the

39

The degree of similarity between the Q-sorts for any two self-images at one point in time is an *intrayear* correlation. There were 28 such correlations every year for every subject. The average of these 28 correlations describes the process of structural integration at a given point in time. Structural integration is one of the two basic processes of identity formation described in Chapter 3.

The degree of similarity between the features of a single self-image in two different years is an *interyear* correlation. Each subject had eight such correlations for any given time interval. The average of the subject's interyear correlations for a time interval describes the process of *temporal stability* for that interval. This is the second of the two basic processes of identity formation described in the last chapter.

The results and analyses in the next section pertain to *groups* of subjects. The two racial groups are compared in terms of specific correlations as well as the average correlations. (A comparison might, for example, involve average values of many subjects and thereby include multiple single correlations.)

Thus, reviewing here the operational meaning of the identity measures and the general types of comparisons to be made serves to emphasize the complexity of these data. This complexity is inherent in the *Q-sort* instrument and, moreover, is desirable. As opposed to a single "objective" self-image study, this instrument is intended to analyze multiple self-images as they integrate at any one point in time and as they change over time. Identity formation is a complex phenomenon which refers to issues of integration and continuity, issues which are at varying levels of individual awareness.[4]

Reviewing the sets of self-images in the first part of the chapter allows for observation of selected areas in identity formation. Processes of integration and continuity occur in each of these areas. When we then turn to the average correlations, it is the processes themselves that become foremost in our attention.

In both types of observation, the primary interest rests upon interracial comparisons. Do the black and white adolescents show different forms of identity formation? And, if there are differences, what do they consist of?

idea of individualized decks for each subject, stems from Prelinger as described in NIMH Grant Proposal M-3642. His contribution to the basic methodology of this study obviously cannot be overestimated.

[4] Among the many arguments for the utility of the Q-sort is that through its many cards not merely the most prominent conscious tendencies are tapped. Other patterns, less prone to conscious control, also manifest themselves in the sorts.

DATA AND ANALYSES

Parental Images

One self-image set studied by the Q-sort was that concerning parental figures and the subject's relationship to them. Erikson and others cite the importance of parental identifications to identity formation. Aspects of these parental identifications were explored here through study of the subjects' idealized self-images with reference to these figures.

Each year the subjects were required to sort the features of themselves according to what they would be like if they were a perfect son to their mother (*ma*).[5] Four sorts later, they were requested to sort the same features according to what they would be like if they were a perfect son to their father (*pa*). The subject's first sort, it is remembered, is always with reference to what he is like right now (*me*).

The intrayear correlations obtained refer to the relationships between the individual's present self-image and that which he believes would most suit each of his two parents. In addition to this, another relevant intrayear correlation concerns the coherence between the *ma* and *pa* images; in other words, how closely does that ideal self-image for father resemble that for mother?

Changes over time were also studied. There are first of all the variations over time in the intrayear correlations among these images. The other measure of temporal change is that of variations in the content of the images over differing time intervals, the interyear correlations.

Results

In comparing the two racial groups, the blacks showed consistently higher correlations than the whites between *me* and *pa* at significant levels (.05) for Years 1 and 2.[6] In later years the correlations for the two groups become more similar, but the values persist as higher for Negroes. In Figure 1,[7] both higher values and less overall change for the Negroes is apparent.

Going on to the relationship between *me* and *ma*, we again find higher correlations for the black adolescents. Here, however, the gap becomes smaller over time as seen in Figure 2. By Year 4 the Negro and white boys

[5] Symbols in parentheses following the description of a given self-image refer to the notation used in the text, graphs, and tables.

[6] See Table 1 in Appendix A. The Mann-Whitney U Test was used to estimate this significance level.

[7] It should be noted that medians are the points used for the curves.

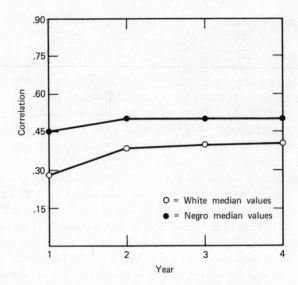

Figure 1. *me/pa* Intrayear correlations.

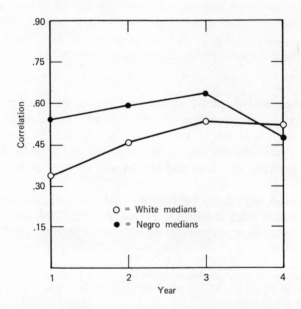

Figure 2. *me/ma* Intrayear correlations.

have almost identical median correlations. This is in contrast with the *pa* image, in which the Negroes maintain a consistently higher correlation with their *me* image than do their white counterparts (Figure 1).

Still another view of the Negro high coherence and low change rate can be gained in comparing the groups for coherence between *ma* and *pa* (Figure 3). Here again the Negroes show a higher and flatter curve than the whites, but in the fourth year the Negroes show a sudden decline in the correlation between *ma* and *pa*.[8] It is the whites who develop a gradually increasing coherence between *ma* and *pa*, a steady increasing integration of the two self-images. The blacks, however, change less for the first three years and then show a sudden decline in coherence as they are about to leave high school.

In Figure 4, by way of triangular representation, the relationships for *ma, pa,* and *me* are depicted. The distance between two self-image points in a triangle is directly proportional to the size of the correlation between them. Most striking here is how the yearly triangles remain virtually congruent for the Negroes. The white triangles, however, progressively increase in size over the first three years. The distinction between the

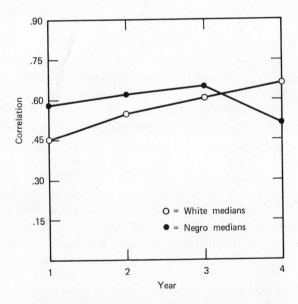

Figure 3. *ma/pa* Intrayear correlations.

[8] Use of the Mann-Whitney test also suggests this reversal; see Table 1 in Appendix A.

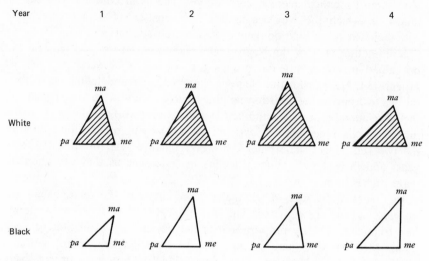

Figure 4. Parental images: intrayear correlations. *Note*: Length of line is proportional to size of correlation between the two self-images involved.

blacks' stasis in parental self-image relationships and the whites' gradual change thus receives further illustration.

Additional evidence of the blacks' *lack of change* is in the content variation studies. It is remembered that if the sorts for specific self-images are compared between different years one can obtain a quantitative estimate of the degree of continuity in content arrangement. Figure 5 displays these interyear correlations for the *pa* images. For *pa* the Negroes have ranges with higher upper limits and higher medians in all but one comparison.[9] There is a similar pattern for *ma*, but for this image the disparity between Negro and white medians is not nearly as great as that for *pa*. Compared with the whites, the blacks have a higher stability of content for both parental images; this quality is especially marked for *pa*.

The dual trends of high coherence and lack of change recur again and again for the Negroes in this parental data. The Negro *ma* and *pa* self-images are, except in Year 4, more highly correlated than the same pair for the whites. And the coherence as well as the content of the images changes less for the Negroes each year. Such a combination of trends is reminiscent of the operational definition of identity foreclosure given

[9] As can be seen in Table 2 in Appendix A, these higher median values show up as positive tendencies in all intervals, and as highly significant for the Years 2–4 interval as determined by the Mann-Whitney U Test.

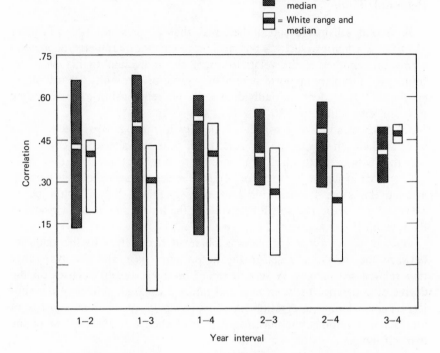

Figure 5. Range and medians of interyear correlations for *pa*.

earlier, that identity formation variant in which intrayear and interyear correlations show little or no change. With regard to the parental self-images, the Negro group manifests patterns that suggest the variant of identity foreclosure.

A further finding of the analysis concerns *differences* between the two parental images: (1) The correlation between *ma* and *me* is always higher than that between *pa* and *me* for the Negroes, but it also fluctuates more than the corresponding *me/pa* correlation. Thus by Year 4 the *me/ma* value is equal for Negroes and whites (Figure 2). (2) In addition to the greater tendency for *ma* to change in its intrayear correlations, its content is also more subject to yearly fluctuation (see Table 2 in Appendix A). (3) Finally, the Negro *ma* image shows higher interyear correlations than does the white *ma*, but the discrepancy between the two racial groups is not nearly as great as that for *pa*. We return in Chapter 5 to consider the significance of these interesting differences between parental self-images.

Personal Time

A second self-image set studied was that of personal time. Primary identification figures, such as parents, represent one parameter of identity formation; another is the relationship of the individual to his past and future. Most directly involved here is the "sense of continuity" that Erikson speaks of.[10] How does the individual sense his relationship to his receding past and emerging future?

In order to study this aspect of identity formation, two specific sorts were used together with the *me* sort. Each year subjects were required to sort the features of themselves according to how they recalled they were "at the start of junior high school" (*past*). A second sort required them to arrange the same features for how they imagined they "would be in 10 years" (*future*). The sort for *future* was the fourth one, that for *past* was the seventh.[11]

One set of intrayear correlations obtained here refers to the relations between the individual's self-image at present (*me*) and the two other time-related self-images. A second set of intrayear results focuses on the degree of coherence between *past* and *future* each year, reflecting the subject's personal time integration. A third group of correlations consists of interyear correlations, emphasizing degree of change in the content of the two self-images.

Results

Comparing the blacks and whites on the relationship between *me* and *future*, we find that the Negroes show striking trends toward higher correlations in the middle years of the study. This is seen in Figure 6, where an initial similarity between Negroes and whites gradually widens; by the fourth year the values have again become similar.[12] While in the process of going through high school, the black adolescents show a very strong resemblance between present and future self-images; but at either end of the process this coherence does not differ significantly from that of the whites.

In reviewing the relations between *me* and *past*, two very interesting trends appear (see Figure 7). First, there is a configuration of rapid change on the part of the Negroes. Over the initial two years they show stable and high values; but then their correlations between *me* and *past* decline rapidly so that by the end of the study the median value is .10, an extremely low

---

[10] See, for example, E. Erikson, *Identity: Youth and Crisis*, 1968, p. 169.

[11] Since *past* refers to a fixed point in time, and *future* to a point always 10 years later, one would expect discrepancies between the two images to increase each year.

[12] Mann-Whitney tests confirm this finding; see Table 3 in Appendix A.

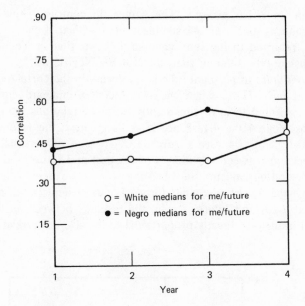

Figure 6. *me/future* Intrayear correlations.

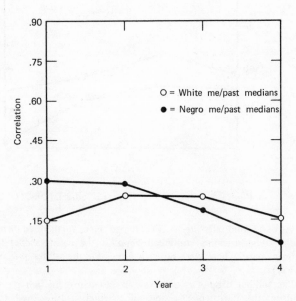

Figure 7. *me/past* Intrayear correlations.

correlation for either group.[13] The whites, in contrast, show a curve of gradual changes and eventual stabilization. By Year 4 the white *me/past* value has returned to the same median value as that in Year 1; this is a result considerably different than that for the Negroes.[14]

A *discontinuity* in personal time also appears in the correlations between *future* and *past*. Here, when the two racial groups are compared, the Negroes are found to have consistently lower correlations than the whites. Finally, by Year 4 the divergence is at its greatest; the median value for the Negroes being .04 (see Figure 8). The sense of personal continuity, of a connection between self-images of the past and future, thus diminishes at an accelerating rate for the blacks.

Through the use of triangular forms, Figure 9 depicts the relationships among *me, past*, and *future*. Again, the distance between each self-image point of a triangle is directly proportional to the size of correlation between

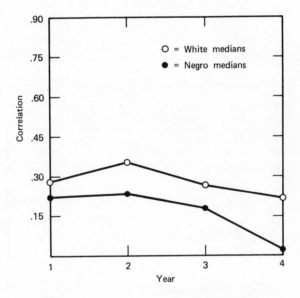

Figure 8. *past/future* Intrayear correlations.

[13] The Wilcoxon and sign tests for change give further confirmation to this graphic decline, demonstrating continued trends to decreasing values for Negroes in later years in *me/past*. There were no such trends for the whites (see Tables 4 and 5 in Appendix A).

[14] A similar sudden drop was observed in discussing the *ma/pa* values for the Negroes, raising the "brittleness" issue, of whether a foreclosed identity variant might be more likely to show tendencies to diffusion.

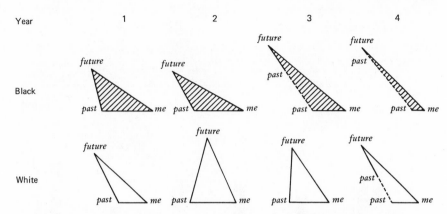

Figure 9. Personal time. *Note*: Length of line is proportional to size of correlation between the two self-images.

the two self-images. Vividly displayed through these figures is the growing *discontinuity*, or "gap," between the black *past* and both *future* as well as current self-image (*me*). Such a consistent and marked trend is not present for the whites.

The same theme of historical discontinuity is reflected in the interyear changes in content (see Figure 10). For the *past* the Negroes show significantly greater discontinuity of content than do the whites. It is of interest that this difference varies directly with the number of years included in a particular interval, for example, in the interval between Years 1 and 4 and Negro median is considerably lower than in the interval between Years 1 and 3.[15]

For the *future* image a Negro discontinuity pattern is no longer present. Now interyear correlations are again higher for blacks and in fact they increase in direct proportion to the number of years in an interval. All black values are high; all medians are above .50. In contrast to the *past*, the Negroes once more show a sameness of content, while the whites have greater lability of content in this image. The fluctuation in content also varies with time for whites. However, the whites show a widening discrepancy in the content of this *future* self-image, as opposed to increasing similarity for the Negroes.[16]

This difference between *past* and *future* can be further illustrated if one compares black *past* and *future* in terms of content change (interyear

[15] See also Table 6 in Appendix A for further documentation of this pattern.

[16] The graph for this set of correlations is again consistent with this. It is not presented here, but can be obtained from the author by the interested reader.

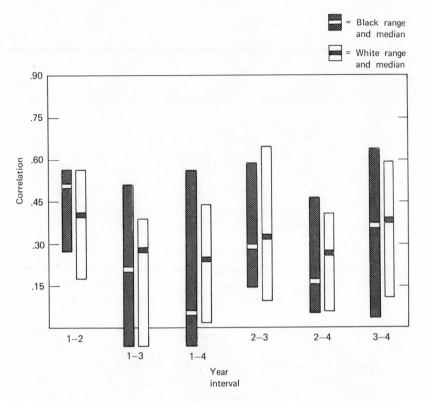

Figure 10. Ranges and medians of interyear correlations for *past*.

correlations) as shown in Figure 11. It is abundantly clear here that the *future* is a more stable self-image than the *past* in *all* intervals; such a difference is absent for the whites.

In reviewing this second self-image set, two important consistencies have been disclosed. The first is really further amplification of the foreclosure trends that emerged in examining the parental findings. For the Negro *future* self-image there is a pattern of sameness of content and high intrayear correlations.[17]

The second consistency appears to be the reverse of the first one. For the *past* the Negroes repeatedly offer evidence of a break—a sudden decline in relevant correlations. This appears in three different contexts: their decrease in *me/past*; their decrease in *past/future*; and, finally, their de-

[17] The averages for *future* reflect the same higher values for the Negroes, see Table 3 in Appendix A.

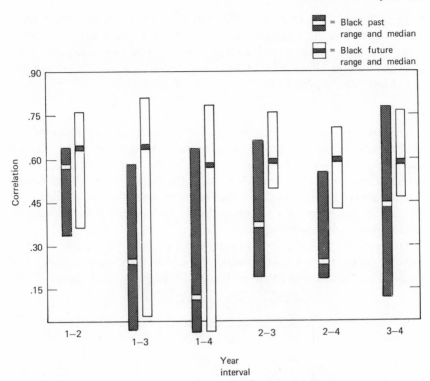

Figure 11. Ranges and medians of interyear correlations for black *past* and *future*.

creasing interyear correlations for the *past*. It is remembered that there was a suggestion of this phenomenon of sudden decline in the preceding section, as Negro *ma/pa* coherence decreased rapidly in Year 4.

At this point there are two different sets of self-images in which tendencies toward variant identity formations are *suggested:* the *foreclosure pattern* for parental images and *future* and the *diffusion pattern* for the *past* and for the *ma/pa* coherence. It is the black subjects who manifest these patterns in all the instances.

The Fantasy Self

A third set of self-images involve the individual's image of himself with a minimum of super ego restrictions, and emphasis on ego ideal. If there were no rules, inhibitions, or barriers, how would he imagine himself to be? What is the *most desirable*, most ideal image the individual holds of himself?

To explore this area the individual subject was asked to sort the features according to how he imagined he would be "if all your dreams were to come true; if everything would work out the way you want it to" (*fantasy*). Both the various vicissitudes of this image over time and its relationship to the individual's present self-image, *me*, were studied.

Results

The findings once again emphasize the rigid unchanging characteristic of black self-images, a recurrent theme in the previous results. The relationship between *me* and *fantasy* shows a consistently higher correlation in all years but one for the Negroes. Moreover, the Negro curve of *me/fantasy* correlations is considerably flatter than the white one as shown in Figure 12. The whites, however, vary in positive and negative directions showing an overall trend to increasing value.

The content variations are similar in form, with the Negro interyear correlations for *fantasy* being consistently higher than are the whites (see Figure 13).[18] Also illustrated in Figure 13 is the related aspect of the

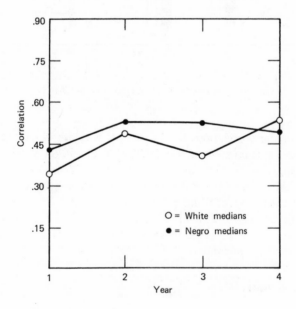

Figure 12. *me/fantasy* Intrayear correlations.

[18] This is confirmed by the Mann-Whitney test as *all* but one of the Negro values show tendencies in this direction and significant results in the Years 2–4 interval as seen in Table 7 in Appendix A.

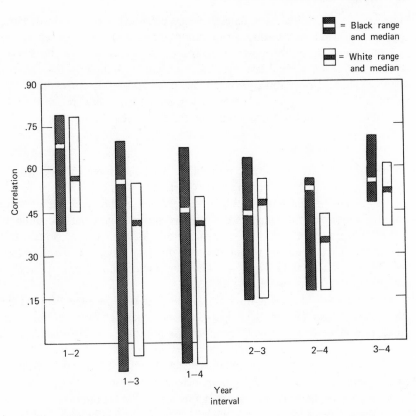

Figure 13. Ranges and medians of interyear correlations for *fantasy*.

whites' greater variation in medians from one interval to the next. Thus the white medians vary from .56 to .35 and generally seem to diminish in value over longer intervals. However, the Negroes show *both* higher medians and less fluctuation in these medians over the entire four-year period.

The black trend toward identity foreclosure receives further emphasis in this third set of self-images. For the Negro boys the *fantasy* self-image is relatively nonmalleable in content. It is also more tightly connected to the self-image *me*.

The Present Self-Image

The present self-image has been referred to in all of the preceding analyses. The relationship between present self-image, *me*, and the other images served as a standard by which patterns of intrayear correlations at single points as well as over time could be observed and compared. However, it is also relevant to discuss the comparative studies of this self-image.

In terms of structural integration aspects, the image *me* can be studied by examining its annual averages of intrayear correlations for the two groups. This average represents the mean of the seven correlations that occur with *me* each year.[19] Temporal variation for *me* is expressed by the time-related changes in the averages and the interyear correlations.

Results

In many respects the image *me* differs from each self-image thus far explored. For the first time there is a virtually identical configuration of values for *both* racial groups. When the medians of intrayear averages for *me* are compared, the values are strikingly similar, with the exception of Year 2 when a suggestion of greater magnitude for the Negroes is present.[20] The curves for *me* are more parallel than for any other set of self-images we have examined (see Figure 14). The Negro curve, however, is somewhat flatter than the white one.[21]

The content variation fiindings also display striking parallels between the two racial groups (Figure 15). Here, the *me* image shows a gradual discontinuity in content between the first two interval comparisons for both groups as manifested by diminishing interyear correlations. Over all later intervals the *me* image shows increasing continuity of content, a reversal of correlations which is true for *both* blacks and whites.

Also present here is the Negro tendency toward sameness, however. In all the intervals from Year 2 onward, the Negro medians are markedly higher than those of the whites. Thus *me* becomes more constant in content for both races, and for the Negroes it remains at even *greater* levels of constancy over these three later intervals.[22] The whites, however, show a

[19] That is, *me/ma, me/friend, me/future, me/pa, me/other, me/past, me/dream.* To obtain this mean, each of the correlation coefficients was first transformed to a *z* value (Fisher transformation) and these converted values were then averaged. All average correlations referred to hereafter in the text were calculated in this way.

[20] This deviation from the pattern here supports the previous noted tendencies of the Negro values. Further statistical confirmation of these results is shown in Table 8 in Appendix A. It should also be noted that if the intrayear averages of other self-images are examined the parallels found here are not present, hence the similarities are not simply a consequence of comparing averages. The groups differ in the same manner as in the specific intrayear correlations observed in the previous sections.

[21] Both the suggestion of greater magnitude and the flatness are consistent with the recurrent findings for the Negro subjects: high intrayear correlations and low change rate.

[22] See also Table 9 in Appendix A for further statistical confirmation.

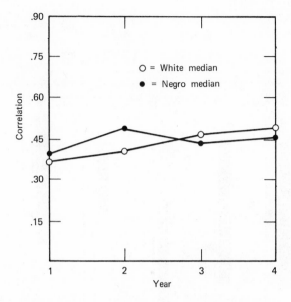

Figure 14. Average *me* intrayear correlations.

gradual steplike rise in these interyear correlations until by the Years 3–4 interval their median approaches a level similar to that of the Negroes.

It is for the self-image *me* that the two groups are least differentiated from one another. However, they are *not* identical. There are subtle indications of differences between blacks and whites, again in the direction of foreclosure patterns for the Negro boys.

The Averages: Indices of Structural Integration and Temporal Stability

To measure the degree of self-image integration for each subject, an average of all his intrayear correlations was obtained each year. To determine the degree of self-image content stability over any time interval for each subject, the average of all his interyear correlations for that period was obtained. It is remembered that these two averages are, respectively, the primary indices of *structural integration* and *temporal stability*. In the preceding sections we considered various facets of identity formation by way of reviewing specific sets of self-images. Now, in shifting to analysis of these averages, we directly confront the overall processes of identity formation. In other words, we now return to the basic question, Do the two racial groups differ in their identity formation?

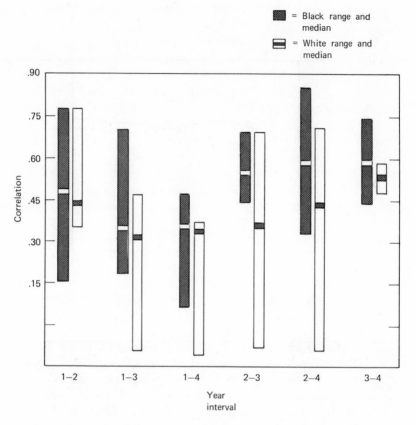

■ = Black range and median

▤ = White range and median

Figure 15. Ranges and medians of interyear correlations for *me*.

Results

The average of each subject's correlations for any given year was calculated from his 28 separate intrayear correlations for that year. The absolute value of this average quantified the fit of all self-images for that year. A value of 1.0 means that all the subject's self-images were exactly the same in that year, while a result of 0 represents absolutely no resemblance of the images to one another in that year.[23]

---

[23] One further possibility exists, and that is a negative average correlation, which would represent a reversal of content. Such an average never emerged, although there were individual correlations of this nature, as is discussed later. Related to this is the fact that an average of 0 could represent a combination of negative and positive correlations. This would also mean that the *overall* fit of images was at a 0-level

It is this average intrayear correlation that operationally defines a basic process of identity formation, namely, structural integration (see Figure 16). Change in the value of the average over time is a reflection of changes in the direction and rate of structural integration. The two racial groups clearly differ from one another in terms of this process. The black intrayear averages show remarkably little change from year to year. Figure 16 reveals the rather flat curve for the Negroes. The Wilcoxon test for change shows *no* significant Negro changes in this average over any time interval.[24]

The whites, however, display a steadily rising structural integration curve. Moreover, use of the sign test on their values suggests change over four different intervals.[25] Still further evidence of these racial differences in structural integration is given by application of the Mann-Whitney test, which compares yearly sets of average intrayear correlations. For the whites three different intervals show the later year as having either signifi-

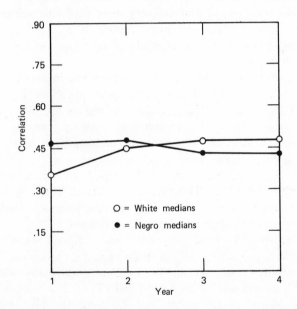

Figure 16. Average intrayear correlations: structural integration.

despite the wide variability in specific self-image partitions. See footnote 19 for comment on the calculation of these averages.

[24] The sign test, however, does suggest increasing values in two intervals; see Table 10 in Appendix A.

[25] See Table 10 in Appendix A.

cantly higher averages or trends in that direction. The blacks, in contrast, show *no differences* within any interval in terms of these averages.[26]

In addition, the black adolescents have a somewhat higher degree of structural integration in the early years. However, this black-white difference is rendered slight when one compares the patterns of *change* in structural integration. With this comparison the racial groups become differentiated from each other. For the blacks the process of structural integration is a static one; yearly structural integration values are essentially without change. In direct contrast, the whites display a process of *progressive increase* in structural integration.[27]

The process of temporal stability reflects this same pattern. Once again, the Negro subjects show a striking *absence* of change. In all time intervals the blacks have higher interyear correlations that do whites. In other words, the overall *content* of self-images between years is emphatically more constant for the Negroes than it is for the whites. Figure 17 and Table 12 in Appendix A document this phenomenon both graphically and quantitatively. The content of their self-images is relatively fixed for the Negroes. However, the whites have unmistakably more malleable self-images. This greater flexibility in white self-image content is reflected in their lower interyear averages and in their generally broader range of average correlations in each interval.

To summarize, it is the blacks' relentless sameness—the "flat line" quality—that is forcefully described by these important averages. Their structural integration, a process earlier defined as the average intrayear correlation, shows minimal variation from year to year. In addition, at several points in time, the actual degree of integration is higher for the blacks.

The other basic process of identity formation is temporal stability. This is the variable defined by the average of interyear correlations. Its value was *always* higher for the blacks. Over all time intervals, then, the black subjects have greater constancy of self-image content.

The values for structural integration and temporal stability are both lower and show a greater tendency to fluctuate for the whites. In terms of identity formation variants, the blacks clearly exemplify identity foreclosure. This variant was earlier operationally defined as being present when "either the structural integration and temporal stability of self-images remain stable; or only temporal stability shows continued increase, while structural integration is unchanging."[28]

[26] See Tables 10 and 11 in Appendix A.

[27] See Figure 16; Table 11 in Appendix A gives statistical indications of this difference as well.

[28] The fact that this variant, as well as the other identity variants, is defined in terms of comparisons over time is discussed in Chapter 3, especially p. 36.

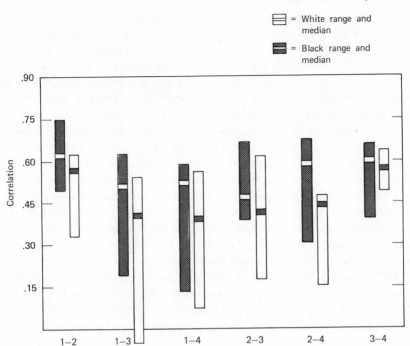

Figure 17. Ranges and medians of average interyear correlations: temporal stability.

Of the two identity formation processes, it is that of temporal stability in which the blacks differ most from the whites. The black subjects clearly show greater constancy of self-image content over time during *every* possible year interval. Additional suggestion of the blacks' static content was given in examining the results for the separate sets of self-images. For each image except *past*, the black interyear correlations were uniformly higher than the whites. This difference is found for the *me* image as well, a self-image in which the racial groups differed less than in any other. The overall black pattern, when both identity formation processes are considered, does represent identity foreclosure. However, it remains an unanswered question as to what the significance is of temporal stability being that process which so emphatically reflects the blacks' static, fixed quality.[29]

[29] A new study, currently under way, proposes to investigate specifically cognitive dimensions of black and white adolescents in addition to identity formation processes. One of its foci will be that of determining whether or not separable cognitive functions are being measured through each of the identity formation processes. A second way one might conceive of this difference in identity formation processes is in terms of

The whites exemplify an overall pattern most consistent with that of progressive identity formation, in which "both the structural integration and the temporal stability of . . . self-images are simultaneously increasing over any given period of time, though not necessarily at the same rate."[30] It is remembered that from Years 2 to 4 there is a gradual increase in the magnitude of white temporal stability (Figure 17). There is also a steady increase in white structural integration from Years 1 to 4 (Figure 16).

These trends are in the same direction as those discussed in the earlier sections, in which specific sets of self-images were explored. In those explorations we found strong suggestion of foreclosure patterns for the blacks and progressive identity formation for the whites. The significance and basis for these black and white patterns in separate self-images and in identity formation remains to be considered. Also of importance are the clinical questions. Namely, what, if any, are the relationships between these quantitative measures and the clinical data, the longitudinally collected case history material from each subject over the four years.

The next three chapters deal with the issues of interpretation, etiology, and clinical relevance. Chapter 5 examines the questions with respect to the sets of self-images. In Chapters 6 and 7 the roots of black identity foreclosure are sought and relevant explanatory models are constructed.

---

the blacks having less opportunity in adolescence to modify their self-images, for reasons to be discussed in the following chapters. However, the internal arrangements and rearrangements of self-images may be less contingent upon the external millieu of figures and institutions and thereby more prone to change, if even minimally. These are the arrangements that are measured by the structural integration variable. Both of these speculations are at this point unsupported by further data. Finding this variation in *degree of* stasis is intriguing and serves to highlight the fact that the racial differences require further empirical investigation.

[30] See footnote 28 regarding the nature of this definition.

CHAPTER 5

Interpretations and Clinical Parallels

INTRODUCTION

The next three chapters are closely related in aim. Together they take on the task of comprehending the many and diverse findings presented in Chapter 4. Detailed analysis and interpretation of these complex results is approached in these chapters, beginning first with the data at hand, then concluding with theoretical models and implications.

In this chapter aspects of specific self-images are discussed in terms of clinical parallels as well as theoretical speculations. A similar approach is then applied to reviewing the identity formation processes themselves. Finally, the intriguing problem of how clinical events articulate with such quantitative findings is raised. In response to this important issue of clinical relevance, the concluding sections of the chapter follow the identity formation of a black and white adolescent through both historical and *Q*-sort data.

PARENTS AND IDEALIZED FIGURES

For the black subjects an important result appeared in the analysis of parental self-images. With respect to both parental self-images, the blacks show relatively unchanging and high correlations with the other self-images. In addition, both images have a relatively fixed content over time. Such trends emphasize the significance of parental figures in *both* the structural integration and temporal stability of black identity formation. There are also some interesting differences between the two self-images. The subject's ideal self for father is a most inflexible image: its correlations show but meager change in terms of both intrayear and interyear values. The ideal self for mother has generally high correlations with other self-images; but it also shows evidence of greater flexibility with respect both to relations with other self-images and in shifts of its own content. Because of this greater tendency to change on the part of the *ma* image, the initial rather high coherence between the two parental images undergoes a decline in the last years of the study for the Negroes.

61

Seen alongside the clinical facts, the results for the black *pa* self-image are puzzling. Almost two-fifths of the blacks had no father at home;[1] most of the others spoke of their father as a disparaged, uninspiring man. However, on the *Q*-sorts the blacks have extremely high and unchanging correlations for the *pa* image. These correlations—both intrayear and interyear-are among the most rigid of the black values. One solution to this puzzle may lie in the following speculation: Given an absent and/or degraded father, his place in the identity formation of his son is that of a rigid unchanging figure whose expectations are imagined by the son to always be the same.[2] In such a situation, unavailable to the son is the chance to anticipate, compare, and modify the ideals he *imagines* are held by his father with his father's *actual* ideals. Through an immense lack of information, the paternal self-image becomes a foreclosed one, a "prepackaged" one, and thereby a major contributor to the identity foreclosure of the son. This kind of sequence is discussed further in Chapter 6.

The interplay between parental images and other self-images for a subject was often instructive. At times, a shift in the intrayear correlations for *ma* and *pa* was paralleled by an increase or decrease in the corresponding values for the subject's *me* image. On other occasions the interaction was with parental images, future, and overall structural integration. Consider the two black subjects, Lenny and Jerome. For these adolescents the correlations between *me* and the two parental self-images significantly reversed their values over a one-year period. The *me/ma* value now became significantly higher than *me/pa* for both boys.[3] This change paralleled decreases in the magnitude of the *me/future* correlation and in the structural integration correlation. Clear historical events paralleled these quantitative shifts:

> *Lenny* had been "persuaded" to go to college by his mother and guidance counselor. To be a Marine as was his father was suddenly becoming an unacceptable ambition in the eyes of his elders, and in part for him.

> *Jerome* was with his mother's urging becoming increasingly committed to a college education and the career of a physician. Continuous sources of anxiety lay in the facts of his dubious academic performance and in his father "staying out" of this decision. Father was a postal clerk. Mother, about to receive her night school degree in education, clearly knew better about these matters.

---

[1] As compared to the whites, one-tenth of whom had no father at home.

[2] This speculation receives support by the fact that when blacks with fathers were compared with blacks without fathers the latter group demonstrated consistently *higher* intrayear and interyear correlations for the *pa* self-image. See Tables 13 and 14 in Appendix A for a summary of the Mann-Whitney tests on these values.

[3] At $> .05$ probability level as determined by the *t* test.

Table 1 and Figure 1 display the correlational patterns associated with these shifts in the valuation of mother and father in the lives of these two boys.

Table 1. Shifting Intrayear Parental Correlations for Two Black Adolescents

Subject and Historical Event	Year	Intrayear Correlations		Structural Integration
		me/ma	*me/pa*	
Lenny	1	.39	.63	.39
Increasing college	2	.69	.61	.44
pressure; anti-	3	.65	.44	.31
Marine ambitions	4	.68	.59	.36
from mother and				
school				
Jerome	1	.69	.70	.54
Increased medical	2	.53	.46	.48
school wishes				
coupled with poor				
academic perfor-				
mance and greater				
urging from				
mother to medical				
school ambitions				

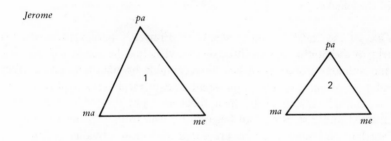

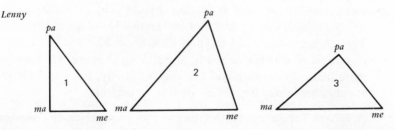

Figure 1. Intrayear parental self-images: two black adolescents. The distance between points of a triangle is directly proportional to the correlation between the two self-images. The number within a triangle indicates the year of the Q-sort.

CONTINUITIES AND DISCONTINUITIES
IN PERSONAL TIME

Relationships between personal time and identity formation were studied by means of the self-images for *future* and *past*. Findings here contained both a parallel and marked deviation from the parental results. For the *future* high intrayear and interyear correlations were repeatedly found in the black subjects. This self-image always differed from that of the whites, and in the same direction, that of *greater rigidity* in terms of both its relationship to other self-images and its content repetition.

However, the *past* self-image disclosed a new pattern for the blacks. For these subjects there was now a persistently low interyear correlation, thus indicating a marked *discontinuity* of content. In addition, the Negroes always showed lower correlations of the *past* with the current self, *me*. Extending the impression of diminished intrayear correlations were the values for Negro *future* and *past* coherence. These values were also strikingly lower than those of the whites. This set of trends is remarkable in that it represents a total departure from all other findings for the Negroes.

Some clues as to the significance of deviation can be found in the clinical data of the blacks:

> One of the most dramatic events in Frankie's adolescence was his trip to the South. It was through this visit that he made the startling and unhappy discovery of how unacceptable his ancestors and current relatives were. The disappointment and chagrin were aggravated by the multiple rejections he then received from these "farmers," who disliked and distrusted him because he was from "up the road" (the North). Although Frankie presented the most obvious example, a similar problem was observed in each of the Negro subjects. The sense of discontinuity with the past of their family and, more broadly, race was present at all times in varying degrees. The predominant pattern was the subject's ready denial of *any* knowledge of his family's history. The subject would claim to know little about his father's background, regardless of whether or not he were living at home. He knew much more about his mother and her present family, but had little to say about either family beyond his parents' generation.
>
> A second expression of this theme was in the Negroes' derogatory comments about the "southerners," the Negroes who were "stupid"

and "out of it." Half of the Negro subjects were born in the South, and all had relatives still living there.

> There was much to indicate shame of all aspects of their past. Benny, at once annoyed and embarrassed, directly criticized his ancestors for being "slaves" and "putting up with it." Another expression of the Negroes' rejection of their past was in their violent dislike of the Black Muslims. Besides urging segregation, the Muslims by their elaborate mythology emphasized the issues of racial history. The subjects viewed the movement with distaste: they were troubled by the "stories" the Muslims told which had no connection to "reality."

The white subjects gave much evidence of a sense of continuity with their past:

> All of the whites knew the history of at least one previous generation on both their mother and father's sides. Some knew of several sets of grandparents, of their life and work in Europe and the United States. There was neither reluctance nor uneasiness in speaking about these events and figures of the past. Indeed, several white subjects were extremely proud of their relatives, admiring their possessions and reputations. The disagreeable elements that must have existed were greatly overshadowed by the subjects' proud memories and abundant factual knowledge of the past. Most importantly, the negative aspects of their histories were not of sufficient weight to even suggest any disruption or denial of a past.

A number of determinants underlie the discontinuity expressed here by the blacks in both the quantitative and qualitative data. One, already alluded to in many ways, is their shame of the past. A second is the actual instability in the Negro families. Many of the Negro subjects had minimal acquaintance with their father and his family. Although these same subjects appear to have greater overall knowledge of their mothers, it is questionable as to how much contact these women maintain with members of the preceding generations. The white subjects have emerged from relatively stable paternal and maternal lineages, have changed residence much less frequently than the blacks, and are often physically surrounded by relatives of several generations. Some have spent most of their earlier years in the context of a large extended family within the same house.

Associated with the phenomenon of black historical discontinuity is the issue of black self-hatred. This issue represents a major source of the Negro subject's shame of his past. It includes the history of slavery, exploitation, rejection, and paternal desertion. It includes the personal history of being "blood" relatives to a series of undistinguished often degraded men. Thus not only were our subjects members of a generally ostracized and disrespected minority, but they were also among the "undesirables" within that group: *men*. The Negro boys at times spoke eloquently of this self-hatred. Although dark-skinned and of southern vintage, they insisted upon their distaste for dark-colored Negroes and for "stupid" southern Negroes.

> In one boy, Benny, the expression was even more direct. To Benny, most Negroes were "coons," "cats who have been 'bad' and drank too much." The Negroes confirmed for him "the image" that society had of him and his people: of lazy, drunk, "no-good men."

These observations of the rejection of the past and the associated self-hate coincide in part with the findings of Kardiner and Ovessey[4] in their emphasis on self-hate. However, there is a lack of support for their conclusion regarding the Negroes' desire for "whiteness," a wish Kardiner and Ovessey suggest as strongly associated with the self-hate. Such a wish was not discernible in the extended interviews, projective materials, or Q-sort data. The impression was rather one of varying degrees of self-rejection coupled with self-resignation.[5] Wishes for who they wanted to be are most closely related to the dreams and wishes of the subjects, to their inner heroes. This area was studied through the Q-sort by use of the image for *fantasy*. Although the results do not confront the issue of "whiteness," they do speak rather directly to the question of resignation.

FANTASY AND FAILURE

Analysis of the black patterns for *fantasy* image disclosed trends already noted for parental as well as *future* images. Intrayear correlations of *fantasy* image with current self, *me*, are consistently high and unchanging. The measures of content change, the interyear correlations, indicate that the *fantasy* image itself is considerably less malleable for Negro boys than it is for whites.

Again, it is valuable to turn to the clinical data for some clues to the understanding of why the Negroes display here a *fixed* image both in terms of structural relationships and content:

---

[4] A. Kardiner and L. Ovessey, *Mark of Oppression,* 1951.

[5] The gamut of implications for self-degradation and self-hate are discussed more fully from the standpoint of an explanatory model in Chapters 6 and 7.

Frankie, a Negro subject, had few positive conscious or preconscious identifications. There were fleeting glimpses in the interviews of his brother-in-law, uncle, and southern grandfather, and even these men were said to have many faults. Most of the Negroes frankly stated that they had no heroes, no people they wanted to resemble now or at any other time. They wished "just to be myself." Occasionally, uncles, teachers, prominent Negro athletes, and Negro small business owners were selected as ideal figures. Negro responses were not as sparse when the question turned to antiheroes, those people whom they wished above all not to resemble. Readily listed were bums, beggars, gangsters, and thieves. In some cases, such as Frankie's, the most prominent antimodel was father.

There were interesting variations on this hero-antihero pattern:

Lenny frequently wished to "take after" his father, to be a Marine. However, the community and his mother seemed to intervene as they urged him to richer "opportunity" and "challenge." Thus one of the few Negroes who found a positive hero in his life, his father, was forced to reject this man in favor of new, and to him unknown, images of Negro successes.

Jerome wanted for many years to be a "doctor." Despite an unremarkable academic record and difficult school adjustment problems, this wish remained strong. His father, a postal clerk, was rarely mentioned in any of the interviews. When Jerome did speak of him it was always in an uneasy embarrassed manner.

Benny was a third variant. He openly and repeatedly berated his "mean" and "stubborn" father. This man, separated from the family for many years, was seen by Benny as an outcast, an undesirable. Benny insisted that the only reasons for any visits to the man were for "cash."

Given what appears to be so great an impoverishment of positive models, it does not seem surprising that the Negro self-image of what he would be like if all dreams were realized is a remarkably *static* one. It is static in relation to both its content over time and its relation to the current self-image. There is but little choice available. Yet the ingredients for fantasies are plentiful. It may well be that the ease with which dreams can become nightmares for these subjects is sufficient to greatly restrict their flexibility in considering *any fantasy* self-image.

There is still a third possible reason for the restricted nature of Negro *fantasy* image. Once more, it is suggested in the clinical material:

> The status of mothers for the Negroes is quite different. Unanimously, mothers were described with the strongest superlatives. They were advisers, exceedingly nurturant providers, and "bosses." Any problems, from the most trivial to the most profound, could be solved by them. They might also initiate major conflicts for their sons in their authority roles. Subjects often worried about their "moms" and their recent report card. To speak of heroes for the Negro boys may, then, be inaccurate. A major figure of conscious admiration and virtual worship was a *heroine*, mother.

One can only speculate as to how threatening such a "hero" may be for the Negroes. One way to deal with the threat of a woman as an ideal image is to tenaciously hold to a consciously acceptable ideal fantasied self-image and to permit absolutely no variation in it.[6]

The white subjects revealed much flexibility in intrayear relationships and malleability in content for *fantasy*. Accompanying this opposite quantitative pattern is a very different clinical picture as well:

> There were an overwhelming number of heroes for the whites. The men they admired varied from history teacher to astronaut. Not only was there an abundance of such men, but the number continuously grew. Almost yearly, these boys discovered new people they wished to emulate either by direct acquaintance or via the mass media. The Negroes, in contrast, only occasionally found a new model. Most often their new ideal figure was a neighborhood bachelor or football star. And within a brief span of time this added hero would be "dropped" from the already small Negro repertoire.

> Another feature of the whites' heroes was the seeming influence that they had. Two white subjects' changes in career plans occurred during and immediately after a year with new teachers of whom they were exceedingly fond. Most convincing, in terms of influence, was the fact that each of their new goals was to become *exactly* what this new hero was. A third subject became increasingly motivated to attend college just after several athletic heroes decided to continue their education rather than enter professional athletics.

[6] Here then is the notion that rigidity in a self-image may offer a form of insulation, protection, from conflicts inherent in the image. This is similar to the hypothesis that identity foreclosure serves such a function for the subject in general as discussed in Chapter 3, p. 30. Other speculations about functions of identity foreclosure are discussed in Chapters 6 and 7.

In addition to greater numbers of positive models for the whites, the nature of their antiheroes also differed:

> There were usually specific people whom the white boys wished to avoid resembling. Other than the generally disliked "garbage man," all other figures were associated more with the individual subject and his tastes than with being white. Joey, for instance, was unique in his hate of the "Boston Strangler" and Oswald. Others chose thieves, murderers, cheats, bigots, and "idle men." There were few antiheroes for each subject. In fact, some subjects claimed to be unaware of anyone they did not wish to resemble. In addition to being small, the individual list generally remained stable over the years. Detested figures were rarely added or deleted.

Using the dual assets of greater availability of positive models and fewer negative models, antiheroes, the whites maintain a *fantasy* self-image capable of numerous changes. Absent for them is both the impoverishment of models and the need for insulation through rigidity.

THE CURRENT SELF: INTERRACIAL PARALLELS AND IMPLICATIONS

Qualities of rigidity and unchanging content are no longer present to any noticeable extent for the current self-image, *me*. Curves of intrayear correlations for *me* over the four years are virtually identical for both racial groups. Not only are the changes in intrayear correlations very similar for blacks and whites, but the magnitudes of the correlations are also near equivalent. Moreover, the content variation findings show many parallels; *me* becomes more constant in content for *both* races.

These findings are surprising and extremely interesting. For black and white subjects, the current self-image undergoes very similar vicissitudes in its correlations with the other self-images and in its own content. There are but undertones of the black *me* having greater constancy of content and stronger similarities to their other self-images. Why then does this self-image show so little differentiation between the two racial groups?

In relation to parental ideals, personal history, and fantasy, the black boy shows trends suggestive of identity foreclosure. For his present self-image this configuration is only subtly hinted at, however.

There are several ways to consider this departure. A number of highly problematic issues are involved in the self-image sets already considered. Among such issues are disturbed parental identifications, with the black usually having an absent or degraded father; an uneasy and largely unwanted relationship of the black to his past; a restricted and conflictful perception of his future; and a fantasy life again influenced by restricted alternatives and the multiple anxieties attendant upon these restrictions. When not brought into direct focus, these disturbing issues are not highlighted in the data. Yet they all obviously impinge upon the subject's present image of himself, and this impingement, this influence, is revealed in many subtle ways as the *me* image is carefully examined. However, when the subject's attention is forced upon the problem areas, their importance and consequences become most clear, as the preceding repeating patterns for parents, personal time, and fantasy have so well demonstrated.

One may, with good reason, question why the impact of these issues upon the black boys is not reflected more clearly by the current self-image. For it is in this image that the subjects appear most similar to one another. The present self-image is the one of which the individual is most conscious and is also the most amenable to the influences of everyday events and stresses in his life. What he *consciously* "thinks of himself" may indeed be quite similar from moment to moment for the black and white youth.[7] Both sets of boys attend high school, have varying groups of friends, go to parties, and have "similar opinions of themselves." When such a highly self-conscious self-image is requested, the possibility of less conscious conflicts being tapped by the Q-sort and thus leading to distinct types of self-image patterns, is minimized.

The other self-images are likely to be less familiar to the subject. The images are constructs he has minimally, or perhaps never, consciously entertained. Being less familiar, they are also less subject to conventional responses, those responses that would be made by many adolescents. Without this shield of conventionality, the various self-image differences between racial groups become more readily apparent. In brief, requesting the unfamiliar runs counter to the option of a stereotyped response, a response cleansed of idiosyncratic shading or conflict. In situations of extreme pathology, for instance, marked rigidity or diffusion, one might expect that even this self-image would reflect such patterns. With rare exception, however, neither of our groups of subjects included such extremes.

Given these assumptions as to what the *me* self-image represents, two important methodological facts become very clear. First, differences in the two racial populations are not to be found in examination of *only* their

[7] This situation may be much different now than in 1962–1967 when these studies were performed. If indeed there are now marked differences in conscious self-image between the two groups, then on the basis of this hypothesis we would expect marked differences in *me* between Negro and white subjects if they were now tested. Such a replication study has already begun.

current self-image.[8] Second, identity formation cannot be studied by means of a single self-image, *me*.[9] This position was insisted upon in the original definition of identity formation. The concept refers to specific ego developmental processes, both conscious and unconscious, involving a wide array of intrapsychic and psychosocial issues, and these issues range from those concerning libidinal needs to significant defenses to community recognition. Such an array of variables is surely not encompassed by the current, highly conscious self-image. At times Erikson discusses the relationship of concepts such as self and self-image to identity.[10] The position taken here is that self-image, as the term is ordinarily used—namely to refer to the current highly conscious image of one's self—is *not* a sufficient index of identity formation. However, it is also not irrelevant; this image must be studied as one aspect of any given identity formation.

In the analysis here the *me* self-image has been used as a form of "base line," an image to which all other self-images have been compared in viewing their intrayear correlations. It is in the interactions of *me* with these other self-images that less conscious, more conflicted issues have become apparent. Through such data, utilizing the other self-images, differences between the racial groups have emerged with increasing clarity.

THE PROCESSES OF IDENTITY FORMATION

No one set of self-images represents a person's identity formation although it may reflect different facets. It is the overall patterning of self-images that defines the identity formation for an individual. The two basic processes of identity formation are structural integration and temporal stability, operationally defined by two specific averages: those respectively of the intrayear and interyear correlations.

The black and white groups diverge in respect to their interyear and intrayear correlation averages. The black intrayear averages are highly

[8] This fact is of course confirmed here. It also helps to explain why Rosenberg did *not* find any differences between racial groups in their "self-images" (personal communication); see M. Rosenberg, *The Adolescent Self-Image*, 1965.

[9] Erikson makes this point in his most recent collection of papers: "[social scientists] . . . sometimes attempt to achieve greater specificity by making such terms as 'identity crisis,' 'self-identity' . . . fit whatever more measurable item they are investigating at a given time . . . they try to treat these terms as matters of social roles, personal traits, or conscious self-images, shunning the less manageable more sinister—which often also means the more vital—implications of the concept," *Youth: Identity and Crisis*, 1968, p. 16.

[10] See footnote 9 and Erikson (1959).

similar from year to year; there are no significant changes in this value over any time interval for them. In contrast, the whites' intrayear averages rise steadily. Over several intervals there is evidence of highly significant changes. In addition to their lesser rate of change, the black averages are of greater magnitude. In sum, the findings are consistent with a rather static process of structural integration for the Negroes. Taken at any separate point in time, the Negroes show a higher degree of structural integration than the whites.

Interyear averages, indices of temporal stability, repeat this pattern. The Negroes have higher interyear averages in all time intervals. Since interyear correlations measure shifts of self-image content, this finding means that the Negro subjects differ from the whites by their maintaining a greater constancy of content for their overall set of images. Once again, it is the Negroes who show an *absence of change*, a stasis. Their self-images are distinguished by the tendency to repeat the same patterns of characteristics from year to year, thus resulting in high interyear correlations. The whites tend to vary in the yearly composition of their self-images, their interyear averages always being lower than those of the Negroes.

The blacks, then, are distinct from the whites in terms of both basic identity formation processes. The black temporal stability is consistently high over the four-year period, and structural integration, the second basic process, fluctuates significantly less for them. Taken together, this pattern of basic processes conforms precisely to that of identity foreclosure. This identity variant is at the opposite pole from a moratorium, in which the individual appears in flux, "finding himself," and "getting his feet on the ground." Rather than any form of fluctuation, the blacks show signs of having a clearly defined array of self-images. From at least the beginning of high school, these images have been unchanging in terms of both their interrelationships and their specific contents.

In many ways this static quality was articulated in the clinical data. It was very striking when black boys spoke of their "future":

> The Negro boys had little if anything to say about the future. Generally, their alternatives were to work in a factory or perhaps join the armed forces. Or, if their dreams were fulfilled, they might some day own a small store and finally be "boss." Frankie, for instance, saw the future only unenthusiastically. He might have a job. However, just as likely were the possibilities of unemployment and a bad marriage. The sensed absence of choice, of "potentials" and "challenges," was more than obvious.[11]

[11] In the next chapter several other clinical illustrations of "stasis" qualities are given.

The whites, in having a lower degree of temporal stability and a fluctuating structural integration, suggested at several points the possibility of a psychosocial moratorium. Moreover, one white subject showed unequivocal evidence of such a state in his marked decline in temporal stability accompanied by a pattern of stable structural integration. Another white boy strongly hinted at a moratorium as his structural integration alternately rose and declined over the years. Yet, the general white pattern was most consistent with that of progressive identity formation, in which both the temporal stability and structural integration progressively increase in magnitude over time.

THE IDENTITY FORMATION OF TWO ADOLESCENTS

Both black and white subjects were presented with several unanticipated changes in their plans for future work and education. Neither the white nor black boys were oblivious to the surprise, nor able to rigidly insulate themselves from its impact. Yet their responses were distinctive, and illustrate in greater detail some differences in identity formation. Two boys, already briefly referred to, provide excellent examples of a psychosocial moratorium and an attenuated diffusion. Here we look more closely at their development and its association with Q-sort findings.

> *Jim, a white, Italian boy.* Jim decided at the close of a year in trade school that this was neither the kind of training nor future work that he wanted. However, having begun the school, and with some parental pressure, he unhappily remained there. He continued to "try it," hoping that welding would become "better." By the end of the second year there he was surer than ever that he wanted to go to college and eventually become a history teacher, most certainly not a printer. To the dismay and chagrin of his family, he began investigations into preparatory schools and colleges. The inner confusion and changes of this period are well represented on the Q-sort (Figure 2). Most dramatically, *all* self-images changed strikingly in content between the second and third years of high school. The greatest change in content was between the first and third years. At times the change was a complete reversal of priorities, as expressed by negative interyear correlations. Two sets of his interyear values are seen in Figure 2.
>
> The negative correlations, indicating the reversed self-image contents, persisted for certain images until the last two years. By the third and fourth years of the study, Jim had become most settled with his decision for prep school and a career of history teacher. This newly

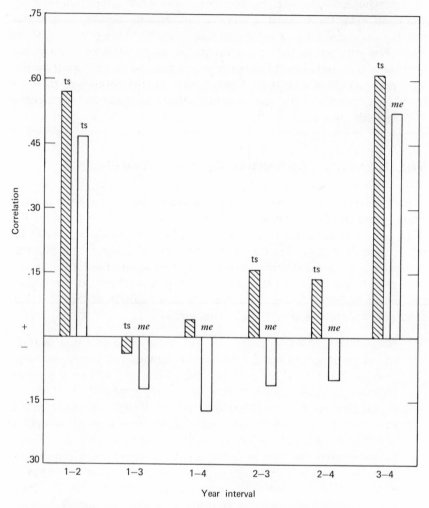

Figure 2. Interyear correlations for Jim: temporal stability.

found certainty is reflected in the set of correlations for the interval Years 3 to 4. In this interval all correlations now return to highly positive values.[12]

---

[12] Had the interyear correlations—particularly their average—remained negative until the end of the four years, Jim would have exemplified our single clear case

Jim remained overtly "stable" despite the turmoil in his plans and the major shifts in the content of his self-images. This clinical presentation was also expressed in his *Q*-sort results (Figure 3). Jim's average intrayear correlations for *future* and *me* declined slightly in the first years and then stabilized or increased in the last two years of the study. His average intrayear correlations, the index of his structural integration, showed the same cycle of variation, as seen in Figure 3. Thus there was a decrease in the degree of structural integration between Years 1 and 2. However, when the content of the self-images was shifting drastically, the degree of structural integration was already

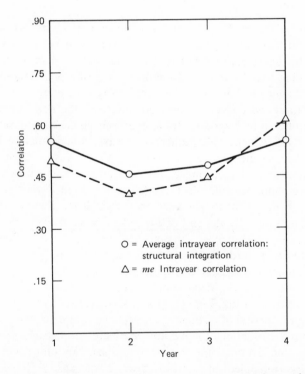

Figure 3.   Intrayear correlations for Jim: structural integration.

of negative identity. The persistence of negative interyear correlations denotes persistent *rejection* of previously esteemed features. Also of interest here is the fact that Jim's *me* shows marked variations from the patterns of the other subjects. Rather than displaying no differences, the self-image here has patterns that parallel those of Jim's other self-images. This deviation is consistent with the hypothesis suggested earlier in discussing *me*, namely, that in extreme situations even this highly conscious self-image shifts as well.

increasing in Years 3 and 4. Such a combination of fluctuating temporal stability coupled with fluctuating structural integration fits the operational definition of a psychosocial moratorium.

The outcome of Jim's moratorium is progressive identity formation. This outcome appears in both sets of data on Jim. Clinically, it is apparent that his new "choices" are emerging as he now speaks of the future and goes about making his very new plans for it. The more objective and quantitative indication that this "experimentation" period has ended is in the Q-sort results. Jim's final Q-sort, taken at the close of high school, displays rises in interyear and intrayear correlations, a finding consonant with further progressive identity formation.[13]

Lenny, a Negro boy. Lenny also encountered unexpected changes while in high school. He had always wanted to be a Marine, as had been his uncles and his father. He would often try on their uniforms, imagining that he too would soon be wearing one. However, in his second year of high school he began to receive increasing encouragement to attend college, to take advantage of the new "opportunities" being offered to Negroes. His mother and guidance counselor echoed these "suggestions." His father, as always, "stayed out of it." Lenny remarked to me that a "reverse prejudice" was occurring. He felt that he was now receiving undeserved attention and support because of race, "not myself." Yet gradually he decided that perhaps the many encouragements stemmed in part at least from his own ability and "promise." By the end of his second year of high school, he had decided to go to college. Nonetheless, he felt that the Marines were not wholly discarded as an alternative. One could go to college and be a failure too, he observed, in speaking of the "college graduates who hung out in the playground" without work.

In contrast to Jim, Lenny's Q-sort showed rises in temporal stability during this period of changing plans for the future, as seen in Figure 4. Interyear correlations for *me* and *ma* increased between Years 2 and 4, and the average interyear correlations, the indices of temporal stability, remained high from Year 2 onward.[14] Hence a major shift in plans and long-range goals was only in very small part reflected by temporal stability measures. The relentless constancy in image content noted for the Negroes as a group was also evident for Lenny.

Lenny's new decisions and orientations were expressed in other

[13] See Year 4, Figure 3 and Years 3–4 interval, Figure 2.

[14] Only on two self-images, *pa* and *other*, did he show any significant decline in interyear correlations.

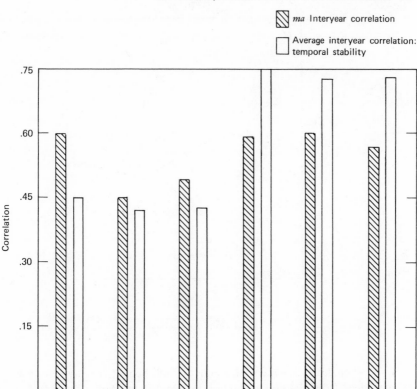

Figure 4. Interyear correlations for Lenny: temporal stability.

ways on the Q-sort: by significant decline in his structural integration values. As illustrated in Figure 5, his average intrayear correlations decreased between Years 2 and 3; his index for structural integration for Year 3 was significantly lower than for Year 2.[15] Thus his response to the newly emerging plans and concomitant changes was one of attenuated diffusion, in which temporal stability remains constant or increases in the face of declining structural integration.[16]

---

[15] At $> .03$ significance level as determined by the t test. To apply the latter technique, the product-moment correlations were first converted to z values. The relevant z averages were then compared using the t test.

[16] One might argue that Lenny's decision, seemingly "forced" upon him, makes his change a radically different one from Jim's and to compare the two then makes little

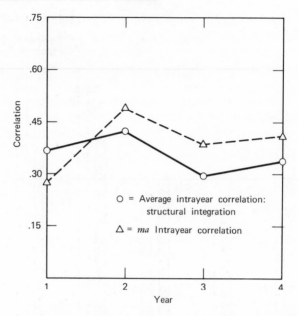

Figure 5. Intrayear correlations for Lenny: structural integration.

When last seen, near the end of high school, Lenny was waiting to hear from the colleges. He had decided to apply, but had done it with a slowness and almost total lack of enthusiasm. He claimed to not know if he could "get in" and he was anything but eager to find out. In addition to the rejection concern, however, there also his clear ambivalence regarding what he wanted to do.

What was absent at this point, was the earlier discomfort and confusion over not following the "plan" he had long ago so carefully mapped out. The situation was now one of passive compliance, with but an inkling of hope on his part that the current plans would fail. For he might then be drafted into the Marines, and thereby return to his originally charted and yearned for course.

The wavering of decisions, the vacillating between envisioned roles of student or Marine, now began to suggest the possible start of a moratorium period. There was no indication of diminished synthesis,

sense. However, this difference in "origin" of the decision change, even if it is significant, in no way explains why in the face of pressure for change Lenny demonstrates so striking a structural integration decrease, a declining synthetic function in face of a rather constant content.

of diffusion of efforts, or of poor functioning. Indeed, much of the earlier intense anxiety had now dissipated.

The Q-sort reflects this clinical absence of diffusion processes. Structural integration has stabilized by Year 4, but the temporal stability values have continued at very high levels. Although not yet clinically apparent, it may be that Lenny is handling the concomitant conflicts over this major set of changes through a return to fixed and highly related self-images.

Clearly, neither of these all-too-brief cases fully explains the many differences and parallels in the identity processes of these two boys. The developments are obviously complex. At this point these subjects have been introduced to exemplify the two predominant modes of identity responses to major *changes* in envisioned plans: by moratorium, which was the more characteristic white response; and by a tendency toward diffusion, a suggestion of a possibly unstable quality inherent in foreclosed identity, the characteristic Negro configuration.

We are left with the puzzling problem of *why* the Negroes show this particular identity formation configuration. Some clinical and environmental clues to the black patterns have already been offered. However, the striking differences between the two racial groups here is by no means explained. In the next chapters the questions of determinants are explored through both clinical and theoretical perspectives.

CHAPTER 6

The Roots of Identity Foreclosure:
Clinical Considerations

INTRODUCTION

By means of the Q-sort and extended interviewing over time, the identity formation of a group of Negro and white lower socioeconomic class adolescent boys was investigated. From its inception and throughout the study, both methodological and theoretical problems were emphasized. Methodologically, a major purpose of the research was the formulation of a definition of identity development which would render the concept amenable to objective empirical analysis. Theoretically, the research was designed to explore the interface between personality development and the sociocultural context. To approach both of these general goals, matched samples of lower-class adolescents, differing only in terms of race, were studied. The results of these studies have methodological and empirical implications.

The first set of results concern methods. An operational definition of identity formation was constructed. This definition was based on the many discussions of Erikson[1] and of other students of identity problems.[2] From this basic definition several variants of ego identity development were given empirical definitions.

This type of result fulfills one of the dual purposes of the study. However, it also generates new problems, specifically, those concerning the *validity* of the operational definitions. Does reformulation of the concept into terms of the complex interactions of self-images over time accurately reflect the meaning intended in prior clinical and theoretical discussions?[3]

[1] Erikson (1946, 1950, 1956, 1968).

[2] See Dignan (1965); Keniston (1959); Lynd (1958); Prelinger (1958); Strauss (1959); Wheelis (1958).

[3] Erikson suggests that identity formation can have a "self-aspect" and an "ego aspect": "One can speak of ego identity when one discusses the ego's synthesizing power in light of its central psychosocial function, and of self-identity when the

One means of responding to this difficult and crucial question was suggested in the previous chapter. There, simultaneous clinical and Q-sort data for two subjects were compared for several periods of time. The conclusions urged by the Q-sort data were examined in light of the clinical evidence of identity processes. Further comparison of clinical assessment of identity development with Q-sort assessments for the same subject should be of value in approaching the problem of validity. Others have suggested means of making and comparing clinical judgments about identity development.[4] The issue here is to utilize such schema in evaluating the significance of the Q-sort technique as it has now been used to study identity development. This evaluation must obviously be made if the technique is to be employed to both define and study identity formation.

The second set of results is a consequence of applying the operational definitions of identity formation. Extensive study of the formal aspects of identity development of all the subjects revealed unanticipated and striking differences between the two racial groups. The Negro boys displayed an unremitting pattern of identity foreclosure. When presented with pressures to change, these same boys responded with tendencies toward identity diffusion. In contrast, the white subjects expressed patterns consistent with progressive ego identity development, and when confronted with unanticipated pressures to change, the whites showed tendencies toward a psychosocial moratorium.

Just as the methodological results generated new problems, so too do these empirical findings. Now the problems become those of explanation. To put it simply: How are we to understand these striking racial differences in identity development? Of particular interest is the identity foreclosure of the black adolescents. This variant is one with crucial consequences for continued ego development. Self-limitation and *stasis* are the basic general properties that define this form of identity development. Rather than manifesting change with further experience and education, the Negro patterns emphasized fixed self-images, unchanging in their content or integration with one another.[5] There are disturbing implications that follow from this variant in identity formation. Minimized, for example, are the

integration of the individual's self- and social role images are under discussion." (1968, p. 211). In these terms we have studied the *self-identity* aspect of identity formation, one of the dual aspects of the overall process.

[4] K. Keniston, "Exploratory Research on Identity," 1959, unpublished mimeo.

[5] The *content* (temporal stability) of the self-images was that aspect most markedly fixed for the black subjects, as noted in Chapter 4. The significance of this difference in degree of stasis between temporal stability and structural integration for the Negro boys is yet unclear. Possible cognitive implications are currently under investigation.

possibilities for modifications in self-image integration, in future orientations, in adaptations to unexpected occupational, educational, or social opportunities. And these forms of constriction represent but a few of the problems raised by this identity variant.

The consequences, both personal and social, of the black identity foreclosure are not in themselves a focus here. However, they cannot be taken lightly. To be sure, they are important as they stress the kinds of serious problems attendant upon such an arrest in ego development.

The discovery of Negro identity foreclosure leads as well to theoretically perplexing questions. Without further study of larger samples, it is unclear as to how general the pattern is. Its significant prominence in this sample alone opens the problem of explanation, however. What, for instance, are the conditions that underlie the emergence of this identity variant *only* among the Negro boys?

Because of these multiple issues, of both theory and clinical consequence, the problem of Negro identity foreclosure is considered in this and the following chapter.

To deal with this identity variant there are two useful and available resources. Most readily at hand are the multiple interviews in which the boys describe themselves, their inner lives, their points of environmental success, and "soreness."[6] Some of this data has already been alluded to in the preceding chapter in attempting to understand specific trends in the two racial groups. We now return to the interviews in the service of inquiring into the nature of black identity foreclosure.

In the next and final chapter, the second resource becomes central. Introduced are the formulations and observations of other social scientists. To what extent do these analyses contribute to an understanding of what underlies the Negro identity variant, identity foreclosure? Two explanatory models constructed from these other accounts are examined closely for their relevance to the quantitative and qualitative findings of this developmental study.

In these concluding chapters, then, the major thrust is toward delineating the roots of the black identity variant. We first consider more detailed qualitative data about the lives of Negro and white boys, particularly their adolescence. Then, two models for understanding the Negro identity foreclosure are reviewed. The complex models are studied in light of how they handle the problem of underlying determinants: the roots of Negro identity foreclosure.

[6] See C. Pierce, "Problems of the Negro Adolescent in the Next Decade," in E. Brody (Ed.), *Minority Group Adolescents in the United States*, 1968, p. 33.

Finally, some remaining theoretical and empirical problems are reviewed. These include questions the study has raised, questions the study cannot answer. Paths to resolving such issues are the final subject of the monograph.

THE SUBJECTS DESCRIBE THEMSELVES OVER TIME: THEMES AND ISSUES

Work

Current employment was a topic mentioned frequently and with disappointment by the Negroes. Being less fortunate than the white boys in finding steady work, the Negro boys held many and varied jobs, ranging from nursery school helper to grocery clerk to subscription salesman. The latter job was particularly despised by one black because of the fact that people "slammed doors at you." When they found work, the blacks were most often discontented with it, feeling overworked or not fully respected. Such problems were never, to be sure, discussed with an employer. Instead, the dissatisfied employee suddenly stopped going to work, giving the excuse of "no time" or illness on the rare occasion when he was pressed for a reason. Indicative of the work problem was the fact the few whites whom the Negro subjects were able to criticize during interviews were always their employers. These were the men whom they perceived as cheating, mistrusting, and exploiting them.

Frankie, a Negro, discussed work extensively at every interview. From the beginning to the end of high school, it remained a prominent issue for him:

Work and Its Discouragements

Each of the years we met, and then at several different times, Frankie told me of his "problem":

> I had a problem. I wanted to get a job. I never could get a job. That's all. That's the only thing that worried me, 'cause I always wanted to get a job. Then when you go look for a job you never find one. . . . I didn't never find one, that's all. I knew I wasn't going to find no job, but I looked anyway. . . .

Agencies and individual employers can never find work for Frankie. He persists in trying, answering the many "no's" with more searching. However, even when he finally finds a job it is most dissatisfying:

> I went down to the employment office and they said they would look me up something. But see anything they want me to do I don't want

to do. Like selling magazines. I don't like to sell mags . . . it gets disgusting and tiresome . . . and you meet a lot of people maybe, you get a lot of no's . . . it's just disgusting.

I didn't like the work so I quit . . . they didn't pay me enough and I didn't like delivering groceries anyway . . . I mean I want to get a job that I like to do. I just wasn't no grocery boy, that's all. It wasn't too cool. It was tiresome.

The work is somehow always degrading. Frankie likes being comfortable and cared for. Work that is physically demanding is thereby unacceptable, and employment that has rejections as an intrinsic aspect is unthinkable, for example, peddling subscriptions or other "traveling sales work." Frankie does not have a favorable evalution of himself. Through devices such as sexual bravado, denials of rejections, and grandiose visions of prosperity, he counters much of this self-degradation. However, work considered "low status" cannot be easily denied or distorted. It is a confirmation of his sensed inferiority and is thus unbearable. The only way for Frankie to deal with this unacceptable confirmation is to avoid it, by quitting.

Finding acceptable work is an old issue with Frankie, his friends, and many generations of his family. This history may be why he shows greater insight and familiarity with employment frustrations than virtually any other issue. Aware that he is still a high school student, not among the "bright" students, and black as well, he does not expect any but the least desirable jobs. Such an accurate appraisal does not lessen the repeating irritation and disappointment.

The current experience is also a foreboding of the future. Frankie has several definite desires for later work opportunities, but grave doubts as to their fulfillment:

I don't know exactly what kind of job I want. . . . I want to be a manager of some place, that's all. [Why?] Because my father used to be a manager. . . . Just have to walk around and do a little bit of work every once in a while and do my own hiring and firing and junk like that. It might be kinda hard getting a job in the future. Lots of things might be difficult. . . . Everything is done by a machine. It's going to be kinda easy you know they won't have a lot of people working. . . . they might have machines replacing men and stuff like that . . . and a lot of people will be out of work. . . . I think it might be hard because from what's going on nowadays it going to be kinda hard. [I] think about it only if I can't get a job all the time. They got, ah, they got somebody always replacing me. It's disgusting. It's a disgusting feeling.

Cogent reasons for Frankie's vague future plans, together with additional motives for his interest in the service emerge from this background of work uncertainty and unfulfillment. To fail in providing satisfactory work opportunities is the community's final insult to Frankie. He tries to deny his immense displeasure: "If I can't get one [a job] it's tough." His many direct and indirect admissions of broodings over this problem quickly repudiate such denials, however. Without any assurance of future employment, and thus usefulness, how can Frankie possibly envision the future with any clarity or security? He can daydream of coming wealth only to be rudely reminded of its unlikeliness:

> I dreamed I had me a lot of money. Had me a tough car. Had me a sweet-looking girl friend. Had me a whole bunch of clothes. I had everything I ever wanted. I had a nice house, I mean those things that will never come true.

Being accepted in the service would mean that those outside of family and friends would not totally reject Frankie, and that in the military he would at least "have something to do." Moreover, he can imagine all manner of advancement, formal recognition, and other specific signs of success. However, again his appraisal of reality qualifies these dreams:

> I had one good dream: I was in the service and I passed the test for lieutenant. That was good and before you know it I had been there two years, and they made me admiral I think. I don't know, a whole bunch of crazy things that don't make sense.

The white subjects spoke less often about their part-time work. When they did discuss this, the comments were usually favorable. For instance, Joey, a white boy, has now worked at the same store for two and a half years. In this period he has been promoted "to the floor" and received several pay raises. Although occasionally critical of his "bosses," his attitude is usually one of immense appreciation and at times frank admiration. A second white subject has had three jobs, each for almost a year. Except for one, all have been in his field of vocational interest, photography. Some of the white boys worked only during summers and were then pleased with the jobs, although occasionally bored. Most important in the whites' descriptions and evaluations of work was the absence of disappointment, degradation, or disrespect inherent in any given job. Moreover, unemployment for the white subjects was not a major catastrophe. Although they were ever alert for a good job, they did not become preoccupied with the problems of unemployment, with the need for "coins" and a "boss." Since the size of family and overall income of both groups of subjects are similar, it is unlikely that this difference in work themes is merely a reflection of economic deprivation. However, it is obviously not wholly separate from financial considerations.

In their plans and expectations regarding future work, the two groups were again widely divergent. The difference here is well characterized in the comparison between Frankie and Joey. Frankie envisioned a limited number of possibilities for future employment, but none of these was anticipated by him as being particularly desirable. Not only was he un-excited about the choices, he was fully aware of their pitfalls. The risk of unemployment in even these jobs was extremely high.

Joey, a white boy, had many plans for his future work. The plans were optimistic and numerous: "I want to get into some good business, make a lot of money, have everything turn out: good wife, kids, give them all I'm capable of."

Of the entire group of Negroes, only two viewed the future as including an array of alternatives and "opportunities." However, both of these exceptions qualified their visions with grave self-doubts as to their chances of success at any of their choices. Hence although they differed from the other Negroes in their conscious perceptions of the future, their final expectations were roughly the same: unexciting, undesirable, but not wholly intolerable work.

These two Negro boys were "deviants" in several ways. One of them, Lenny, was briefly described in the last chapter. With great uncertainty, he eventually acquiesced to the urgings of his mother and teachers and applied for admission to college. Gradually, his ambition and dream of becoming a Marine was tempered. Almost simultaneously, Lenny's belief in his ultimate failure became more explicit and constant. He had seen many "college guys" in the school yard loafing all day because "they couldn't make it." It was perfectly clear to Lenny that they had wasted their time and money in going to college. He was hardly convinced that his future would be any different. Benny, the second "deviant," was one of the most unusual boys in the whole sample. From the start he was remarkably inquisitive, labile, and imaginative. His desires in both the present and future were diverse and elevated, varying from laboratory technician to Ph.D. in philosophy. However, he too questioned the attainment of *any* of these ambitions. Indeed, he eventually almost fulfilled his underlying predictions of failure as he dropped out of school and was repeatedly unsuccessful in enrolling again. When last seen, Benny was disappointed and bitter, working full-time at the very factory jobs he had so adamantly wished never to have.

In place of the virtual sense of *predetermination* expressed by the blacks was the whites' assumption of "free will." Most felt that they could "decide" what their future vocation might be. One had wavered for many years before and during high school. He finally chose to become either a technician or "engineer." The majority of whites considered several alternative sets of plans while in high school. Joey and James typified the pattern of "conversion" from skilled labor to "professional" work. Mel and Paul were subjects who revised training programs and eventual vocations several times. On each occasion they announced a possibility previously unmentioned. In brief, work in the present and future was not seen by the whites as the dismal impasse anticipated by their Negro counterparts.

Heroes

In this thematic area, as in the preceding, the patterns of the Negro and white groups closely parallel those of Frankie and Joey. Frankie had few positive conscious or preconscious identifications. There were fleeting glimpses of his brother-in-law, uncle, and southern grandfather, and even these men were usually said to have many faults.

In place of an idealized figure stood a despised one. The issue is poignantly illustrated as Frankie describes such an "antihero":

"Pops"—A Threat and a Burden

Since childhood, "Pops" has been a feared, generally avoided, object of shame. Dislike—"hate"—of him has intensified, recently coinciding, Frankie thinks, with an increase in his father's drinking. The pattern of Frankie's despair over his father is an old one, however. Were it not for one probable screen memory and an occasional comment Frankie would succeed in presenting a thoroughly bleak view of this man and his relationship with him.

As a child, Frankie remembers that he rarely saw his father. There was a male neighbor who took him and his siblings on trips. They often wished aloud that this man were their father. Nevertheless, Frankie was not completely deprived of his father. Only because "Pops" taught him to "form" did Frankie finally conquer the bully who fought with him each day. This memory of father's guidance is outstanding in its isolation. All other recollections are unpleasant. Among the painful memories are "Pops" losing his "good job" because of drinking, "Pops" as a loud drunkard in the street in view of Frankie's friends, "Pops" openly flirting with street women, "Pops" ' mysterious absences, and Frankie's brothers beating his father at home. Frankie has openly insulted his father's women, although he usually ignores his father in public. In fact, he tries never to be seen with him. This is no great departure from earlier habits. As a child, he felt his brothers were "Pops" ' favorites, and consequently he was unhappy while with his father. Frankie is not alone in avoiding public and private encounters with "Pops." "Moms" does not go anywhere with him and, she confides to Frankie, she is even unhappy in bed with him for he reeks of alcohol. Frankie has made some attempts recently to approach his father, asking him why he continually drinks. "Pops" discourages the attempts, perhaps responding to their provocative character by telling Frankie, "You wouldn't understand." Frankie's angry response to this is, "Later for you."

A puzzling question is why "Pops" remains with the family. He is subject to abuse from Frankie and his wife, who taunt him about his "wealthy" background which did nothing for him and his "brains" which he is wasting. He is otherwise ignored and treated like a "bum."

The motives for "Pops" ' remaining at home are, with only Frankie's view at hand, essentially undiscoverable. Consequences of the nondesertion are not so obscure. By staying home he has taught Frankie to "fight" like a man. Through talk and possibly observations he has shown Frankie that sexual prowess is a critical dimension of manhood. For although Frankie is bitter about "Pops" ' extramarital sexuality, it also represents one of the few ways in which Frankie is proud to resemble "Pops." Encouraging these two essentially masculine similarities represents the positive aspect of Frankie's relationship with his father.

On the negative side are failure, disrespect, deprivation, and much anger. "Pops" is a man "who could have made it." He is not, to Frankie, stupid or incompetent. However, for some unknown reason he has perpetually "messed up." Even with all the advantages of supposed wealth and intelligence he has failed. The unhappy career of his father remains an enigma to Frankie. Yet whatever his confusions about the "why" of his father's fate, the "what" is only too obvious: a fallen man. And whatever failings he cannot himself perceive, he is forever told of by "Moms": the paychecks wasted on drink, her shame of being with her husband, her surprise that he is not yet "sick," and her desire to separate from him. It is unclear whether or not "Moms" expects irresponsibility and failure from all men. Certainly her repeated injunctions to Frankie not to drop out of school, not to get "in trouble," not to marry young, hint at a pessimistic view of *any man*. There are few desirable goals that she depicts for Frankie. Perhaps if he can avoid the pitfalls inherent in being a man he will be told of its potentialities and rewards. Her expectations are reminiscent, though possibly less severe, of those held by some black New Orleans' "matriarchs":

> Men are all alike. Men with six or seven kids at home, sitting at the bar fooling around, spending money. I don't have anything to do with men like that. I like to sit there and watch them. I have no sympathy for them.[7]

[7] Rohrer and Edmonson (1960), p. 129.

"Moms" ' actions and attitudes toward "Pops" have been justified to Frankie in terms of irresponsibility and general unworthiness. Frankie has thus witnessed "Moms" ' managing the family and home, while "Pops" remains the "burden," barely tolerated, heartily disliked, and scorned.

An important, yet still unsolved, problem is whether or not these many years of observations have encouraged Frankie to view men, and thereby himself, as always second in command to women, inevitably less reliable and weaker. While such specific information is not available, we do know much about the closely related problem of Frankie's confused sexual identifications.

Until the age of eight Frankie enjoyed playing with dolls, stopping only when he learned that "boys don't do that." An important element of this play was tearing the dolls apart. Aside from dolls, children of the opposite sex were favored companions. Throughout childhood, and currently, Frankie's happiest times have been with women and girls, whether talking or having sexual adventures with them. His hair has represented a sexual issue. While interested in "processing it," he has rejected the plan, in large part because "it's like going to the beauty parlor." The impetus for a fight Frankie began on the bus came when the opponent "hugged me like a girl." A more explicit homosexual incident evoked a great disgust and violence in Frankie toward his eager seducer. Further sexual uncertainties are revealed in his frequent and lengthy narratives of sexual conquests, describing in elaborate detail virile adventures and successes. Whatever feminine tendencies the above adventures and battles may be responses to are more directly seen in picture-drawing, in which Frankie consistently draws a woman as his first "person."

There are many grounds for Frankie's doubts about his masculinity. In addition to the degraded pictures of his father drawn by "Moms," his experience with "Pops" has been both minimal and displeasing. When with his father he did not feel wanted or liked as much as his two older brothers. Now as an adolescent he is either rejected by his father in the few approaches Frankie makes; or he rejects father verbally—"Later for you,"—or physically when Frankie beats the intoxicated, irritating man.

While contact with "Pops" has thus been limited, and then adverse, experience with "Moms" is of an entirely different nature. Almost all of Frankie's time at home is spent with her and is usually extremely pleasant. If it is not with "Moms" it is with one of several favorite aunts or grandmother, or his older sister. There are an abundance of admired females, but males are few and undesirable: two brothers who "messed up" and a father who continually exemplifies failure. The few more "positive" males are either distant or cautiously admired: a "bank president" uncle in the

South, a Negro football star and, more recently, an uncle who supported and helped provoke several of Frankie's fights during a visit to the South.

In view of this history, it is hardly surprising that Frankie speaks of much anger and resentment toward his father. "I hate him," "I wish he'd get out," are remarks that he makes spontaneously and fervently. Nonetheless the feelings are probably embarrassing as well, at least when expressed to a white adult, for Frankie waited two years before mentioning them in our interviews. There are many ways in which "Pops" has "done Frankie wrong." Though he did not desert the family, as many of Frankie's friends' fathers did, it was almost the same. Feeling deprived of father was a frequent experience. In addition to this sense of abandonment, Frankie witnessed and at times participated in the drunken assaults, which he could actively respond to only many years later. In sum, "Pops" is an inconstant and threatening figure. It is unpredictable when and how long he will be home, whether he will be violent or peaceful, kind or rejecting.

It is of interest that Frankie has not chosen "the gang," and organized delinquency as a solution to the absence of adequate masculine models and symbols. In this respect his nondelinquent older brothers and the occasional yet available exposures to favored uncles have perhaps been important. Furthermore, the mere presence, no matter how unstable, of his father at home has probably been critical in providing an image of manhood as well as tempering distortions of men that "the women" may have created. It is far from certain that Frankie will not choose a path of minor delinquency and petty crime, however. He has on occasion shown interest in this; police have frequently supported images of Frankie as "hoodlum" and tensions with "Pops" have become increasingly unacceptable in the last two years. Homosexuality does not seem to be an acceptable solution to Frankie, and withdrawal from all social contacts is also intolerable, as Frankie shows an ever-present and strong wish to be accepted. One solution, which Frankie himself offers with considerable ambivalence, is "the service." A short period in the armed forces may provide Frankie with both a respite from his immediate family and many new, perhaps to him more acceptable, male models and symbols.

Frankie is a specific instance of the general trend observed in Chapter 5. Most of the Negroes frankly stated that they had no heroes, no one they wanted to resemble now; they wished, "just to be myself." Occasionally uncles, teachers, prominent Negro athletes, and businessmen were selected as ideal figures. Negro responses were plentiful when they considered antiheroes, those people whom they wished above all not to resemble. The subjects then spoke of bums, beggars, drunkards, gangsters, and thieves. For several boys—such as Frankie—the most prominent antimodel was the father. The possibility that at least some of these emphatically detested figures were strong unconscious identifications was suspected for several subjects. The high correlations on the Q-sort between *me* and *pa* for Negroes further supports this speculation.

The Negro subject's deep ambivalence about his father is clearly shown by Jerome. Briefly mentioned in the preceding chapter, Jerome was a Negro boy who for many years had the ambition to become a physician. Despite an unremarkable academic performance and an unhappy school adjustment, this wish remained strong. His father, a postal clerk, was rarely mentioned in any of the interviews. When Jerome did speak of him, it was always in an uneasy, self-conscious way. At the time, Jerome's mother was completing her training at the state teacher's college. She was clearly the principal adviser and motivator in Jerome's thoughts about college and professional training. A related issue for Jerome was his increasing discomfort in remaining in the study. This culminated in his finally telling me that he could no longer come to the interviews since he disliked talking about himself. While in the study he had become highly suspicious of my motives for speaking with him, more than once telling me that we were studying "Negro families" like the studies in his mother's textbooks. Aside from fears of a white, or anyone, peering into a potentially unstable family, it is likely that such discomfort was related to the many sessions with an adult male. He was being forced to speak and introspect with a man—and then a white one—who essentially represented the attributes not found in his father. The conflicts aroused by the study may well have been beyond his capacity or desire to tolerate.

Benny represented another variation on this theme. He openly and repeatedly berated his "stubborn" and "mean" father. The latter, separated from the family for many years, was seen by Benny as an outcast, an undesirable. Benny visits him only for "cash." However, money seems to be in large part a rationalization for the visits. Benny often hesitantly expressed the less critical and affectionate feelings he held toward his father as he recalled the talks and "fun" he had with his "old man."

As was suggested in reviewing Frankie, the status of mothers for the Negroes is quite different. Mothers are always described with the strongest superlatives. Many of these women sounded exactly like Frankie's "Moms." They were advisers, providers, and "bosses." Any problems from the most trivial to the most profound could be solved by them. They had major authority roles in the lives of their sons. Subjects were often worried about "Moms" and her reaction to their behavior, school failure, or unemployment. To speak of heroes for the Negro boys is probably incorrect. They admired, respected and revered a heroine, "Moms."

There were many many heroes for the white subjects. The contrast between Negro and white boys here is richly depicted in comparing Joey and Frankie:

"Fine and Successful": An Abundance of Heroes

Frankie could easily describe those men and boys he wished not to emulate. Finding desirable heroes was far more difficult for him, although when pressed he could recall a few men he admired. For Joey the problem was reversed. He has always been surrounded by relatives and others who were "the best," whom he wanted to "be like." These idealized figures are great inspirations yet, one suspects, always unreachable. The consequence of this was chronic dissatisfaction. It was a dissatisfaction without the element of despair, however, for Joey retains the image of what he can be, if he "tries harder." Frankie, however, expresses a self-dissatisfaction coupled with despair. For Frankie the visions are of what he can and will be if he is not careful. The majority of images of the future for him are foreboding. Only a small number of the men whom he knows or is related to have "made it." The primary problem for Frankie is to avoid becoming like the negative models; the images of those who have "succeeded" are neither vivid nor plentiful ones. All of this, of course, is in terms of conscious or preconscious emulation. There is every reason to believe that unconsciously Frankie has very strong and basic identifications with precisely the men whom he now consciously shuns. These are the men such as his father and others he has known and heard of all his life.

There are but few people whom Joey wishes to avoid resembling. His antiheroes, small in number, all have certain features in common. Their failing is associated either with emotional weaknesses and conflicts, or with work. He does not want to resemble one of his aunts who is a person "with a very bad disposition. She's very slow, takes a long time to get started and gets aggravated if you rush her. She doesn't understand if you rush her." Another kind of emotional problem, far more despicable, is that of rampant aggression and impulsivity:

> The Boston Strangler he killed a lot of people, that guy. If he gets caught boy it's all over. They only kill'em once. But I think it's gonna' be painful, 'cause they gonna' wanna', like the guy that assassinated Kennedy. I'd hate to be in his shoes when he got caught, Oswald . . . . He woulda' died a slow death, Oswald, even if they let him rot in prison. What people would say and what people would do to me if they ever saw me [if I were like him].

> That guy's weird [the Boston Strangler] . . . . Well, when he gets caught he knows he's gonna' die for the one murder he committed. But just think if he had all those lives to live he'd die all those times.

Fifteen times and uh he's gotta' have that on his conscience. He's gotta' think about it once in a while, how he killed the girl. . . . He must have nightmares. Even if he didn't have those nightmares it's just when he's walking down the street or something there's gotta' be somebody that resembles someone, like say somebody has the same dress on that the lady he killed had on. So he just looks at her ya' know and he starts getting nervous and confused. And it brings back memories of what he did, thinking of how he killed and everything.

In Joey's descriptions of these antiheroes are embedded his greatest fears: guilt, death, and violence—about which he is highly conflicted as well. Joey is typical of the white subjects in the importance he gives to conscience and guilt. Although there was variation on this issue with Joey representing one extreme, the entire white group nonetheless differed markedly from the blacks in this respect. For the Negroes the overriding emotions and conflicts centered on rejection and inferiority as illustrated by Frankie.

A second feature shared by Joey's negative heroes is that of degrading work. Here, on the surface, there is a similarity to Frankie, to whom the *wrong* kind of work was a critical problem. However, Joey's objection was almost a fine distinction compared to Frankie's gross rejection of certain jobs for their humiliating features. Joey's concern was about fair compensation and adequate working conditions rather than despicable forms of work. It was a *union demand,* not a demand for *a union* and a *new image:*

[I wouldn't want to be like] my mother now. She's a cashier and has these long hours so sometimes we never know if we're gonna' see her or not. She comes home late at night after leaving early in the morning. I wouldn't want to be like my father in the future because his job has back-breaking hours. He works long hours and takes a lot of money out [for taxes].

And Joey does not want to be like his teacher:

. . . The kids give them a hard time all day. It's a hard job, a lot of kids. They put a bad mark without asking you to be quiet. They suspend the kids. . . .

It is of interest that Joey focuses on work in describing people and life styles he would not like to resemble. Work had undoubtedly been a primary target of the many gripes and bitter comments he had heard from respected adults.

As far as people who are attractive models for Joey, there are more than 30 men whom he admires and in various ways wants to emulate. This number increases each year. Foremost among these virtual idols are Joey's relatives. For great-uncles and a cousin he has unqualified praise.

I wanna' be like my great uncle who helps with the church. He's very famous, very good and well known. He'll do anything he can for people; if he can't he'll have others do it. He helps parish families' juveniles in court . . . They wanted to make a movie about him. He worked his way through college. He's mayor of his village. I want ta' be like my uncle who's in the construction business. He's one of the best masons in the city. He helped build this building. Everybody knows him. He's famous.

His older cousin is a more immediate hero:

My cousin was prom chairman last year and uh he had to turn away a few hundred couples because it was so crowded . . . I guess that was a big responsibility and everyone looked up to him. I asked him if I was elected if he would help me. . . .

The position Joey takes toward his father is more complex. For many years prior to high school Joey wanted to do the same type of work as his father. To his disappointment he was discouraged:

I thought of working near my father as a welder but he talked me out of that. He didn't want me to work those hours. He worked twenty-four hours when the railroad was active. I guess everyone wants to be like their father because they admire him.

Since then he has been critical of his father's work, but also very pleased at each opportunity he has to work with him:

I like working better [than at other jobs] with him [father]. Though it's dirty, it's nice to see the room change from empty to filled with sinks and stuff like working with a man's hands instead of machines. . . . When we were all through he would say "carry the tools back to the car." He would pick up these big pipes. He was all right. He was good to me you know.

Joey has many times enthusiastically told of a summer when he worked with his father, who is a plumber "on the side." He again worked with him after his family moved. This time he was two years older and less patient with the authoritarian regime:

He's [father] funny. If you start something you gotta' finish it. He won't put if off. We screened in our porch, screen windows . . . and I said "but Dad, I'll do it this way." And he said, "You don't want to do it my way?" And I said, "No." He said, "You're going to do it anyways. . . ." Now the windows are all in place. . . . He's the authority.

Despite the apparent problems with this firm ruler, Joey makes it plain through this and subsequent descriptions of their work together that such times are special ones. In fact, it is likely that when working with him Joey is happiest.

When moody, Joey has been told "you are just like your father." In his friendliness and "popularity" at school the resemblance to his father is also clear:

> . . . He comes home in a bad mood or something but my father is very friendly. We used to go out riding with him and we would see somebody and he would go around the block. . . . And he would start talking and talking and they wouldn't stop.

Though he may behave in similar ways, Joey is painfully aware of the fact that he does not physically resemble his father. Joey's shame of his body and short stature is clearer in light of his description of "Pop." "He's sort of big, bigger than you [5'6" interviewer] much . . . . About five ten, 230, 40 maybe [pounds] solid. Big guy."

Increasingly during high school, Joey has looked outside his family for models. He has not, however, to any extent rejected his idols in his family. They are supplemented and "updated," but remain the foundation upon which all later heroes rest. Two recent figures were high school "athlete-scholars," who faced a choice between physical or mental prowess, a decision Joey is also concerned with. At the time they decided to go to college, Joey announced to me that he too wanted to go on to "college." A third of these new heroes was the astronaut, Edward White. What impressed Joey most was his act of rebellion, his refusal to return to the space ship after having stepped outside. Choosing this characteristic is consistent with Joey's tempered forays into rebellion throughout the three years of high school. White's rebellion was not a flagrant violation:

> I think that he [White] was too thrilled to think of getting back in. That's why he didn't think anything of the time that elapsed from the ten minutes he was supposed to be out. . . . I would probably have stayed out as long, if I could've, if the whole United States wasn't yelling at me to get back in (chuckles), I would've stayed out. But he did the right thing. . . . He just proved that man can take that endurance in space up there.

There are many other heroes, ranging from Gary Cooper, to new singing groups, to President Kennedy and the local mayor. Each has something Joey wants to have, be it money, fame, or "ruggedness." Such a profusion of ideals is not confusing, though. Somehow, they are all related to various themes and issues in Joey's current life and do not in any way lead toward severe diffusion of his self-images. They do, however, both because of their number and variety, give increased impetus to Joey's efforts for self-improvement.

In short, the white subjects had numerous heroes, and these heroes appeared to have much influence on the boys' decisions about the future. Joey exemplifies this in his decision for "college." Two other boys decided to become teachers during a year in which they were taught by two men, men whom they were openly fond of and admired.

The Future: Limitations and Successes

The future had very different significance for each subject. His skin color was a major determinant of its significance. The Negro, for instance, saw a future that mirrored the present as he knew it. He might repeat the persistent series of discouragements, or he might live a life resembling the style of adults who now surrounded him, the adults who "never made it." Among these adults were even those who—as Lenny put it—despite "education" were still "on the street corner." The images of storefront drunks and deserting fathers were also, of course, close at hand. It did not require the probing of unconscious fantasies to bring forth images of failure.[8] There are, however, some pictures of the future that are *less dismal* than others. Frankie, for example, imagines a bleak but not wholly despondent period that will follow the completion of high school. His tortured uncertainty and pessimism reflect an outlook and style characteristic of the Negro subjects.

The Future: The Service and Unwanted Marriage

Among the few inspiring figures in Frankie's surroundings have been those in uniforms of the armed services. They possess unequivocal signs of belonging to an honored group of men, at once responsible and important. Frankie is well aware of their influence on his future plans:

> I'm going into the service. I'm going into the Air Force or Marines or one of them. . . . When you see a real man from the Air Force or Marines from anywhere around here, I mean you want to go in it. You say you are going to do it when you grow up, so I just did that when I was little . . . it comes up every once in a while. Then I say, that's where I'm going to go, something like that. I want to wear a uniform 'cause they make you look nice, make you look neat . . . I want to be important.

[8] This is similar to the point made by W. H. Grier and P. M. Cobb in *Black Rage*, 1968, regarding the role of abundant Negro prostitutes and the degraded self-image of one of their female Negro patients.

Frankie has no other plans about the service. For many years he has known he will join after high school. Whenever he reconsiders, however, the idea becomes more tenuous. A particular stumbling block has been the extent of his commitment. The military in part represents to Frankie as it does to several of the subjects a psychosocial moratorium. However, to each of the boys there are specific expectations and anxieties about this desired period. Frankie does not want it to mean a "quitting" or permanent departure from the community, or more important, from his family: "I'm not just going into the service and give up and pass my time away." The specific duration of service is hence a major point of anxiety and confusion:

> I might not want to wait five years, no I might not want to wait ten years. I'll probably wait five years and then get out [of the service]. . . . If I stay there for quite a while, if I stay there for about four or five years and then like it, then I'll tell them I want to go ten years. . . . You might want to get married or something like that when you get out you know. That's what I was thinking, 'cause you might not want to stay in there all your life.

Besides his worry of looking like a "quitter," Frankie worries that the service may rekindle his authority problems: "I might not like getting bossed around. You have to take orders." Yet there are potential immediate gratifications: "Best think I like about it [the service] . . . I guess the food. 'Cause they say you can eat all you want . . . something like that."

All things considered, the military offers a threatening set of problems. He will be with only men and mostly whites at that for much time; he will be away from his home and mother. There are inherent authority as well as probably homosexual conflicts. The pleasure and general reassurance of female companionship will be largely prohibited. Each year Frankie becomes more tentative about the service. It will not be surprising if he abandons the plan to volunteer at the conclusion of high school.

Frankie's thoughts about a future marriage represent rationalizations for shortening or changing his military plans. He rarely speaks spontaneously of marriage aside from discussions about the armed forces. When he does entertain the marriage topic it is in distress, in response to fears of "messing up," being forced to marry. Marriage would consist of a series of cautious compromises with potentially unworthy women:

> I wouldn't do my wife wrong. I would stay home and I would just have to give up a lot of things. I would work and everything. You know support the family and jive. . . . Just as long as she don't do me no wrong or nothing like that I'm going to go along with her . . . do all the important jive that they say when you get married. . . . In

sickness and all that jive. I mean I'll help out. I'll do my part as the husband. . . . Anyhow I'm not thinking about getting married. I like to have fun. If you get married you can have fun, but you have to get out of school and work and all that junk. It's disgusting.

The few pleasant fantasies about marriage emphasize Frankie's intense dependency wishes, now to be fulfilled through wife and money:

I'll get a job before I get married. 'Cause I'm going to have a lot of money saved up anyway. . . . Make arrangements you know. I'll get married and all like that. Just lay up and take it easy for the rest of my life. Ah think me a lot of weird things, boy. I swear I'll make me a good ol' time.

The future and absent heroes have much to do with one another for the black adolescents. Related to the missing heroes and abundant *real* failures are the dismal and tenuous images of the future. There is little—as Frankie so emphasizes—that is desirable and obtainable in the future.

The whites "looked forward" to graduating from high school, to the new opportunities and possibilities they envisioned as now becoming available. For Joey the future had very specific meanings:

The Future: Business and Money

Joey's visions of the future are closely connected with those of work. Rarely does he think of one of these topics apart from the other. The unusual times are when he thinks of money and the future, but even here the link with work is almost a direct one. He has many plans and wishes for future occupations. Both in breadth and in number of conceived possibilities he is once more so strikingly different from Frankie.

Chronologically, Joey first wanted to be an airplane pilot. This was followed by wishes for the priesthood while in junior high school. Soon after, his thoughts turned to the idea of being a welder, "working near my father." Welding and preaching were both discouraged by his father and mother, respectively. He was "talked out of being a priest" because of "too long a schooling." As for welding, his father "talked me out of that; he didn't want me to work those hours he worked when the railroad was active." By the start of high school, having been discouraged from both mentally and physically demanding jobs, Joey had narrowed the choice down to business manager or construction worker. Being a businessman would depend upon college, and the latter was an uncertain but hoped-for goal. What he wanted from his future work incorporated important wishes: "If I have a good position I'd like it, like foreman, manager. It would be good paying and people would look up to you, make you feel like you're a father, ask you for help."

His general plans of business or construction work remained stable through the high school years. Changes within the framework included a refinement of the business category, the category Joey favored anyway. He began to think about IBM training and electronics. However, the goals or "dreams" were the same. Joey did not think that there would ever be reason to alter them:

> . . . make a lot of money, have everything turn out: good wife, kids, give them all I'm capable of. I want to see other parts of the world, of the country. I want the business to be one where I'd have to travel a lot, see how the different businesses are making out.

There were many people who directly influenced Joey's ideas about future work. His father virtually commanded him to become a business man as he ordered him to not take a "hard" job like his: "He said I should be a 'white collar worker.' I told him I'd try to go to college and become an accountant." Others who supported Joey's upwardly mobile plans for the future were his "successful" uncles, and a cousin, a doctor who "helped addicts get off dope."

As opposed to Frankie, Joey described but minimal fears of future failure, of unemployment or of financial indebtedness. In part, this difference is conditioned by Joey's active use of denial. To some extent, even if he were concerned about the future he would be reluctant to reveal this. However, there were a sufficiently high number of interviews and samples of Joey's behavior to have at least suggested some anxiety regarding future failure, if such a latent theme were present. What fears he had concerned the problem of his training after high school. Holding a relatively mediocre pre-high school record, and average high school work, Joey had grave doubts about "making it." To be accepted for further education was important for his current status vis-à-vis his college group and for realization of his "businessman" plans. Conflict over this became increasingly intense as he neared his senior year. His fears of rejection by college suddenly disappeared with an unexpected rise in his grades, however, and what he then read as unequivocal encouragement by his guidance counselor of his ability to pursue further education. Still unsure of his qualifications, this confirmation by his counselor was demanded and obtained.

Further progress in Joey's image as a future businessman was accelerated by achievements in his current job. The most recent promotion placed Joey "on the floor." He is now a cashier, wearing a tie and coat each day and feeling more like a businessman than ever before. The transition from high school student to future manager is already in process. In fact, at this point it might be reasonable to predict that in 20 years Joey will be a "supermarket manager," with a large family and a home in the suburbs, encouraging his son, Joey Jr., to be a professional, such as a doctor or lawyer, not just a "white collar worker."

DEGRADATION AND DIMINISHING INFERIORITY: SELF-IMAGES

Interwoven throughout the preceding narratives—and through the years of interviews—were the Negroes' degraded self-estimates, their unremitting belittlement of themselves. Frequently, the scathing judgments were implicit in other descriptions, in other discouraging episodes. Sometimes the self-opinions were explicit, painfully obvious to the subject and shared with the interviewer.

The themes of worthlessness, undesirability, and uselessness recurred in many contexts. There were no Negroes for whom these themes were subtle. For the whites such topics when present seemed minor: worthlessness did not assume the same unmistakable prominence. A white boy would question his value most often at moments of "self-doubt." Even then the questions would be tempered by a fundamentally optimistic belief that things would always "get better." Once again, the lives of Frankie and Joey reflect these issues as well as the poignant divergence between the black and white adolescent in this area of self-degradation.

One response Frankie had to inferiority was in his "ideology":

An Ideology of Money

While marriage and the military are at most highly tenuous desires, money is the most certain and tenacious of all Frankie's goals. To have money is to be assured of being able to "lay up" (be comfortable). To have money means no longer any sense of inferiority when with those who have "coins":

> Some boys that have money you know they think if you never have no money you know they think you was going to beg them for money. . . . They don't bother me none. . . . If they feel like giving me coins to buy something I mean I don't mind myself. Long as they don't try and say I beg all the time. Which I do not beg.

The magic substance, money, and the material possessions it allows, are the answer to all problems. The multiple fantasies of "success" so constantly drawn by television, radio, film, and the street have obviously had their impact here. However, the intensity and almost complete one-sidedness of Frankie's money wishes suggests that they are something more than reflection of an admittedly dominant cultural theme. To be "secure" and to be upwardly mobile, for example, having "a house in a new neighborhood," are important consequences of having money. Most of all, though, Frankie's envisioned utopia would consist of protection from degraded self-images and from the multiple threats represented by other people. Frankie would be surrounded by things of all kinds and be at once elevated and insulated:

> . . . I'd wish for some money. That's all I would need. If I had some money then I could get me just about anything, I'd want then . . . buy me a new home, buy me a car, buy me some cars, buy a lot of things, buy me some clothes. If that could happen right now I'd wish for that stuff. Buy me a new hi-fi, hmm, buy me a new tape recorder, buy me two or three new tape recorders, four or five amplifiers, three or four record players, all the records I could buy, all the albums I could buy, all the speakers I could buy. Ah, I mean I'd have my house fixed up real sharp. Wouldn't that be a good idea? Buy me some new furniture for myself. Fix up the den.

Occasionally a wife and/or family is attached to the reveries of money, but these people are decidedly peripheral to the immensely gratifying dollar dreams. It is as if such people are added because "you're supposed to." The panacea is money. With it Frankie can cloak all uncertainty and protect himself from all opponents, finally becoming "king."

Other elements of Frankie's ideology are rarely so explicit. The world consists of "haves" and "have-nots" and, less easily expressed, whites and blacks. Frankie's few but significant reflections on "having this kind of hair" and his very conscious experiences of racial rejection together with his awareness of money make it clear that he doubly degraded: he is both black and a "have-not." However, to Frankie, in large measure one can overcome inferiority through riches. The catch is that only by magic or unexpected gifts can these riches be acquired. The solution is not to be found in working; for work presents many problems and pitfalls. It most certainly cannot bring many of the possessions one needs for success.

Aside from money, Frankie is interested in action and "thrills." Women and athletics are major sources of these satisfactions. In general, women are to be exploited financially and sexually. However, there is a limit to the application of this dictum. Frankie frequently feels guilty for "taking advantage" of girls. In fact, one of Frankie's few moral precepts is ultimate respect for females. Toward men such a commitment is not evident. All other of Frankie's moral principles and judgments are apparently derived from the basic commandment, "Never disobey your mother." Trouble with "cops," problems in school, or early marriage would be judged wrong because "Moms" is against them.

Frankie has no formulated opinions or interests in anything that is not of immediate consequence to him. He finds international, national, or community affairs of no importance. As observed above, he is indirectly preoccupied with social class. Racial discrimination is ignored until he is confronted with personal consequences of it. He then responds to the one specific problem at hand. Though living in the midst of active civil rights organizations, Frankie has no wish to either join or learn about them. The current "civil rights" environment has undoubtedly influenced his recent protests against bigotry at school, but any conscious preoccupations about racial issues are for Frankie restricted to the Muslims. Their beliefs in separation, violence, and idolatry are spoofed at and resented by Frankie. He emphatically rejects the movement and denies any interest in it. Moreover, as if suggesting the potential temptation this group has for him, Frankie notes that he will be happy when this group disintegrates.

Frankie has, then, a simple and vague ideology. He has scant concern with his past either in itself or as an influence on his present life. His future is but thinly outlined and then usually only upon request. Ideological devotions and preoccupations are indeed not typified by Frankie. Ideas, conscious beliefs, and abstractions are for others. For Frankie the immediate "kicks" are what count. Even the barest planning or "arrangements" are painful and are only occasionally attempted. However, if ideology qua ideology has not been a crucial concern, the issue of work has been. It is in this realm that society's appraisal of him and his relationship to the community is communicated. These messages—as depicted earlier—have been uniformly discouraging.

Included in themes of work, heroes, and the future have been many indications of Frankie's self-image and its vicissitudes. Though it was not always intended, many of Frankie's tests and interviews have portrayed the panorama of self-evaluations and visions that he presents and in general *dislikes*.

Of Hoodlums and Kings

Frankie has used a minimum of ten names in describing himself over the years of our meetings. The significance of each name, its conscious and unconscious meanings, obviously vary considerably. The mere plentitude of epithets and the richness of images they convey is in itself testimony to Frankie's present and quite possibly lasting conflicts over his self-images. In the following discussion these names or "self-labels" are described and their meanings elaborated.

1. *Good boy.* "I'm a good boy, I think," is a statement usually appended to Frankie's descriptions of battles at school or with the police. It is said

with obvious equivocation, at times with much teasing, as if he were a young boy having just been scolded and now telling of his supposed virtues.

2. *King.* Frankie wants to be "king" and to be called such. It means to be popular and powerful among all people. His popularity is uncertain, always contingent upon performance in school, on the street or, more recently, with white girl friends. The aspect of power is in part one of physical strength and Frankie claims respect for this as he "protects" black sophomores at school. However, the other meaning of power rests on money and prestige. These, as we have seen, are Frankie's dearest longings and the ones he senses are least likely to be attained.

3. *Playboy.* This is a generally valued appellation, with aggressive and masculine overtones. Throughout the years of interviews he has shown preoccupation with "babes." In all relations with them, the goal is to "get them down" (have sexual intercourse). As Frankie repeatedly emphasizes, he has no great difficulty succeeding in this quest:

> I always get involved with a girl. That's the only way I have fun. Yeah man, ah always have me some fun with them. . . . Ah do everything. . . . That babe is crazy about me. I don't know what's wrong with her. I'm a mess, right? . . . I got this strong game. I know how to run a game on a babe, make 'em like you. . . . All the babes love me too much. I'm a mess. I game 'em all down.

His use of the name "playboy" increased considerably in the last two years of the study, possibly reflecting Frankie's increased uncertainty over his own masculinity. Although the reasons for Frankie's greater reference to this name are unclear, the passages leave little doubt of the self-depreciation that accompanies it. It is as if to once again remind all concerned that he is essentially "a mess," in case we forgot while listening to his accomplishments.

4. *Bum.* Frankie says he wants least of all to become a "bum," as are the drunken men sprawled daily on "the avenue." His father in part represents such a person to him, thereby heightening Frankie's fears of this outcome for himself. Although a heavy smoker, he avoids all alcohol, vowing not to "take after Pops" in any way, especially in drinking.

5. *Lazy.* On the road to being a bum is to be lazy, and Frankie frequently thinks of himself as lazy in relation to work and his dislike of most unskilled jobs. In addition to his father, he has his older brother as a model of "laziness."

6. *Hoodlum.* Frankie once remarked about his girl friend's mother who did not like him: "She must have thought I looked like a hood or something . . . asking me all sorts of questions as if I was doing something wrong all the time." Being thought of as a hoodlum by this mother, if the perception

is accurate, continues a tendency begun many years ago by several groups of policemen as they detained and interrogated him for walking on the local university's campus. It is intermittently reactivated. Now this view of Frankie is spreading to older women, as well as to people whom he usually considers "right." The dismay and uncertainty that this change engendered were apparent as Frankie asked me, following his story about his girl's mother, "I don't look like a hoodlum, do I?"

7. *Smart*. This word has two different meanings for Frankie. One is his retort to teachers' and others' low estimate of his intelligence: "I'm intelligent. I'm smart." He needs much support on this point; he usually makes it clear he believes himself to be stupid. In fact, one of the functions the three and one-half years of interviews unwittingly served for Frankie was to encourage him along these lines of hesitantly questioning this belief. He has been able to talk with a professional, a "doctor," and to participate in "research." The other meaning of smart is "wise guy." Especially with men, Frankie is often smart in this way, acting in a provocative and frustrating manner. One of his main aggressive outlets is through "being smart."

8. *Crazy*. During one of our conversations Frankie suddenly told me, "I thought you were going to say I was crazy." He has on other occasions spoken of "seeing things," quickly adding that "I don't see nothing that might make you go crazy." There is good reason for Frankie to entertain such doubts. Episodes of aggressive outbursts have increased recently, disappointments and disturbances with "Pops" are now more acute; rejections when he visited relatives in the South were bewildering and resented. Less easily specified is the diffuse uncertainty Frankie is showing about his manhood and future. There are no data to suggest that he is experiencing any delusions or hallucinations. More likely, it is his diminished control over his aggressions and more rapid mood changes that imply "craziness" to Frankie. It is of course uncertain how adolescent-specific these changes are. There is indeed the possibility that these changes and the ensuing doubts of his sanity will persist throughout an erratic adulthood.

9. *Blue boy*. Recently, Frankie has spoken of racial prejudice and rejections. He usually handles these in the way he treats other very troublesome conflicts: "I try to forget them [those times when discriminated against]. If you think about them too much you go crazy." Sometimes, as in an episode in which he perceived flagrant bigotry in his gym teacher, Frankie directly confronts the antagonist. Then he does not as readily deny the impact of the event or subsequent feelings. The discomfort following the latter incident is still great, and Frankie "tries not to think about it."

Even more troublesome is Frankie's specific skin color which is a dark brown. Much has been written of the status significance of the varying skin colors within the Negro community. Frankie tells several stories about his family perusing their group photographs with guests. Much laughter always ensues because Frankie is "dark," while all the other children, except his older sister, are "light." Frankie, though, emphatically denies that color differences among Negroes are of any importance to him: "It don't bother me." In choice of girls as in self-evaluation, skin color—he insists—is irrelevant. His older brother, foolishly in Frankie's eyes, "likes light better."

To further complicate any inferiority he may feel because of his dark skin is the fact that Frankie's father is also dark. Frankie is preoccupied with not wanting to resemble this degraded man in any possible way. Yet here he must face the most obvious resemblance of all. It is hardly surprising then that Frankie employs fully his much used pattern of denial for this conlict-ridden subject. Although he suspects many personal weaknesses, to him his most critical is that of his color. His siblings for many years could not be stopped from reminding him of his unwanted skin color as they taunted him with "blue boy" (dark skin). The label continues now with white classmates "in jest" shouting "blue boy."

10. *Up the road.* Some of Frankie's most difficult experiences occurred in the South, following sophomore year of high school, when he suddenly realized that he was a northern Negro and estranged in many ways from southern Negroes. This would probably not have been so disturbing were it not that Frankie's immediate ancestors, and oldest sister, were also southerners. The Negroes he met were, on the one hand, "stupid" and "primitive" people to be shunned. On the other hand, they somehow included his own relatives and in part himself. If he were to reject or downgrade southern blacks did that then mean his own origins were lowly and to be ashamed of? The rejections and abuse Frankie received from southern Negro boys make it clear that Frankie was an alien. This left him in an extremely awkward position, for he had not previously thought of himself as a northern Negro. With this new concept, additional dimensions to his self-image emerged. To be "from up the road" means, among other things, to be less like deprived, segregated Negroes. It also—and here is where the confusion lies—means less pure, more like "white." To be similar to the white man is a wish which for Frankie and many other Negroes is fraught with ambivalence. Witness, for instance, the Black Muslims and Frankie's angry protests against them. To learn that he was a northern Negro was probably helpful in terms of self-definition. Yet the confusions generated by the experience were many. The lingering distress of the summer was soon apparent when Frankie assaulted a new boy on the school bus, a southern Negro.

Such an ever-expanding catalogue of self-derision was not evident for the whites. Rather, as Joey's development illustrates, the process was in the opposite direction. The themes were there, but becoming less and less confirmed by both the boy and his surroundings:

An Ever-Diminishing Inferiority

First and formost in Joey's self-image is his body concept: he is "short" and "skinny":

> My arms are skinny, my chest needs development. . . . Everyone says I look old for my age. I think I'm pretty tall now . . . I hope I'm not short at seventeen. I hope I'm big and won't shrink.

Compared to his friends he is "the small one." Joey frequently cites his short stature as a prominent defect, wishing soon that this would change. Whether describing basketball or street fights, he usually refers to his size. Though generally he degrades himself for being short, sometimes the emphasis is on being "tough anyway," or on "not taking it," no matter what the opponent's size. There is something very attractive for Joey about his "Jewish friends' " seeming lack of concern for these dimensions. In addition, even if he is physically lacking, it is his cleverness, the mental qualities that he can assert. The Jews represent in part for Joey a group in which his physical handicaps are not emphasized. In this respect they help him to mask blemishes. The intelligence and education that he credits them as symbolizing are, however, second best; were Joey strong and as physically able as he desired, the "brain" qualities would not be quite as attractive.

However, Joey does not even fully qualify for this "booby prize" group, for his image of his intellect is also not an exalted one: "I don't like to look like a doctor, they're supposed to be real smart and know a lot. I'm not." Until his senior year of high school, Joey felt he could "get by," but he was not going to be very successful academically. "Almost making honors" his last year at school was a genuine surprise. He had studied more assiduously than ever before and most unexpected of all was the suggestion that college would be a feasible choice for him. To the observer Joey's image of himself as "not bright" is less evident than his condemnation of his physical stature. To Joey his outstanding defect is size and puniness; after that, almost as if linked, come his intellectual failings.

What is important in this self-depreciation is that none of these estimates are absolute. On the two standards against which Joey measures himself he received a low rating. However, he can *always improve*. By going to the "Y" for physical development, as he did, he can build up his muscles. By standing up for his rights "on the block" he can counter any impressions of weakness. Intellectually, he continues to demonstrate improvement to himself. The groups he joins are among the "brains" of the school. His grades have risen to the point at which he now has "about the best report card in my home room." He believes in and is committed to *personal progress*.

There are more stable and consistently positive features in Joey's self-image. Reliability, steadiness, honesty, and service are all qualities he admires in several adult figures and tries always to emulate. He believes himself "popular," but never popular enough to be content. The equation is almost quantitative; the greater the number of friends I have—the better I am as a person. Finally, there are the associated qualities of "independence" and dignity to which Joey clings despite his longings for popularity and his often degraded self-image. To a large extent this is in reaction to many dependent trends. Even when his dependency is taken into account, however, Joey's insistence on dignity is not fully explained. In his dealings with employers, friends, and their parents, Joey indicates in no uncertain terms where his limits of compromise are. To yield beyond these lines— whether to please others or achieve further acceptance—is forbidden. The sources of this important theme, pride and independence, are discernible in the models and heroes in Joey's life, and a major source of support for the enhancing self-esteem rests with Joey's milieu as well. The responses of others are uniformly encouraging; promotions, the change from basement to "the floor"; an increasing number of friends from the prestigious Jewish group; "almost making honors"; followed by being told to "apply to college."

CHAPTER 7

The Roots of Identity Foreclosure:
Two Theoretical Models

INTRODUCTION

The purpose of the following discussion is to account for the pattern of Negro identity foreclosure, using in large measure the qualitative thematic data detailed in Chapter 6. To explain the occurrence of this identity variant, two models are considered. The first is addressed to the array of sociological and psychological forces that are in many ways interdependent and closely linked to identity patterns. This model chooses as its focus the interplay between "community"—in its broadest sense—and identity configurations.

The second model is that elaborated by Erik Erikson. Using the framework of epigenetic development, the foreclosure pattern is analyzed with respect to what underlies its prominence within the black sample. In this model development is a central notion. Rather than asking how the identity patterns are maintained and perpetuated, the questions dealt with are those of historical roots: those of the individual life cycle and its multiple relationships to psychosocial events.

THE PSYCHOSOCIAL MATRIX OF
IDENTITY FORECLOSURE

The configuration of identity foreclosure is defined by *stasis*, a marked diminution of change in all facets of multiple self-images. Inherent in this stasis of self-images is the individual's experience—conscious and nonconscious—of the restricted alternatives. Sociocultural and intraphysic components contribute to this experience.

The Environmental Contribution

Proceeding first from the individual's sociocultural milieu, there is the *actual* fact of limited choice. This fact is referred to again and again by Frankie and the other Negro subjects. They notice it most directly in the area of work opportunities. There were generally few part-time and summer jobs available; and those available positions seemed always associated with distasteful kinds of work. White boys found jobs that were, temporarily at least, satisfactory, and within the job there was the opportunity to "progress."

It is not only in the present that work is both limited and unsatisfactory for the Negroes. Prospects of a broader range of jobs after graduating from high school seemed just as unlikely to them. The armed services was one of the few systems in which one of the black adolescents could find some chance of recognition and even—perhaps—advancement. Other than the military system, the blacks saw primarily dull and distasteful factory work ahead of them. At best, there was the possibility of someday being a "boss" of a small store.[1]

A second environmental constraint is in terms of *heroes*. There were few positive adult figures whom the Negro subjects were interested in emulating. Occasionally, popular Negroes such as Joe Louis, Jackie Robinson, Sidney Poitier, and James Baldwin were admired by the black subjects. None of the subjects considered the potent talents of these men as being within their reach, however. As observed earlier, the heroes were not only few in number. In addition, their number gradually diminished over time. Both the size of the list and its continuous diminution stood in marked contrast to the whites' large, ever-expanding list.

Related to both the limitations of work and heroes is the envisioned future. Take a job market that is small and unattractive. Then add few—if any—attractive or appealing adult examples of how a boy might appear in 10 years. Then, as if these facts did not make a dismal enough picture, add restrictions as to where the black man may live and play. To further stigmatize the future there are an abundant number of men who are on the street corners: the unemployed, the drunks, the "bums"; these are men who provide the black adolescents with what Benny called "the Negro image."[2] Such models of the future are seen each day. Even if they are not seen "on the avenue," "Moms" is around to remind her son of his likely fate. In the instance in which his mother is not forecasting such unhappy

---

[1] Not only do the subjects in the study observe these *actual* work limitations; Pierce (1968), Grier and Cobb (1968) and Clark (1965) make similar observations.

[2] See Benny's description of this notion on p. 66.

destinies, frequently the Negro youth learns of his or his close friend's father who has deserted, or who remains and is the object of constant disdain, such as Frankie's "Pops."

From multiple sources the message is clear. For the lower socioeconomic class black adolescent there are few options. Still another message is given by these constraints, that of *devaluation* by the community. Erikson speaks of the importance to the adolescent that he be "recognized" by the community around him:

". . . we speak of the community's response to the young individual's need to be "recognized" by those around him, we mean something beyond a mere recognition of achievement, for it is of great relevance . . . that he be responded to and given function and status as a person whose gradual growth and transformation make sense to those who begin to make sense to him."[3]

Opportunities for new functions and status by way of work are sparse in both present and future. This impoverishment can only lead to the conclusion by the "young individual" of the community's misrecognition—or nonrecognition—of him. Restrictions in living area, recreation, and heroes further inform the Negro adolescent of his inferior standing in the community.

Confirmation of this inferior status is given as well by his personal history and of the history he knows of his race. Each Negro in the study made it clear that he wished not to speak of his own past.[4] Each of them knew more than glimmerings of the broader past of slavery, submission, and ever repeated degradation. Grier and Cobb describe this knowledge of the past:

"History is forgotten. There is little record of the first Africans brought to this country. They were stripped of everything. A calculated cruelty was begun, designed to crush their spirit. After they were settled in the white man's land the malice continued. When slavery ended and large scale physical abuse was discontinued, it was supplanted by different but equally damaging abuse. The cruelty continued unabated in thoughts, feelings, intimidations and occasional lynchings. Black people were consigned a place outside the human family and the whip of the plantation was replaced by the boundaries of the ghetto."[5]

[3] Erikson (1968), p. 156.

[4] The Q-sort results of high negative correlations of Negro boys' present with their past self-image further supports this.

[5] Grier and Cobb (1968), pp. 25–26.

Impacts with the Environment: Intrapsychic Contributions

The message embodied within the complex of restricted alternatives—social devaluation—has several intrapsychic consequences.

To begin with, there is the self-depreciation so poignantly depicted by Frankie and his fellow Negro subjects. Low *self-esteem* is a topic commented upon in every discussion of Negro "personality," both public and not so public ones.[6] Given the many indications he has received of his inferior nature, it is hardly surprising that the black youth assumes they are valid. The category he basically belongs in—"lame," inferior, rejected—is insisted upon in much of his daily experience. The process is continued and amplified when this classification is not only accepted but also in certain ways maintained by the individual and his family.[7]

The same social conditions, limited options, have a second intrapsychic impact. Constant insult, in the form, for instance, of being deprived of the opportunity to earn more money or to have a more esteemed social position, leads to frustration and anger. Until but recently (and primarily after these boys were studied), this anger has been inhibited from overt expression. One well-known consequence of unexpressed anger or aggression is *self-hatred*. Negro self-hatred is a focus of recent analyses by Pouissant and Grier and Cobb.[8] Why this anger has so long been inhibited, finding expression predominantly as self-hatred, is not clear. Rainwater suggests that self-hatred plays an important adaptive function for Negroes in their reaction to "inimical social conditions."[9] Whatever the reasons for the lack of direct expression, the consequences are more than apparent. Replacing overt rage has been guilt, recrimination, and self-hate.

It is important to distinguish self-hate from low self-esteem. In the latter state the issue is, "I'm not good enough; I'm lacking." Where self-hatred is present, the issue becomes, "I don't deserve; I am so despicable that I'm not worthy of this or of any opportunity." Both these states are extremely important in understanding the black adolescent, for they both lead in the very same direction of devaluation and depreciation of self.

In addition to inhibited rage, negative identifications also underly self-hatred. From early in his life until the present, many surrounding familial figures, especially men, have been despised and shunned by the black subject. Sources for this response have come from his immediate family, and

[6] See, for example, Karon (1958), Kardiner (1951), Pettigrew (1964b).

[7] This process is discussed by Rainwater (1966). It is reviewed in much detail later in this section.

[8] See A. Pouissant, "Negro Self-Hatred," *New York Times Sunday Magazine*, August 20, 1967; Grier and Cobb (1968), especially pp. 3–22.

[9] Rainwater (1966), p. 175.

so often from the figures themselves. Frequently, "Pops" is inconsistent, giving out great sums of money and advice early in the day, while later or perhaps the next day becoming vicious and no longer interested. Or "Pops" may be unpredictably moody as he once again is unemployed and in no great demand. Another common pattern is that of father, or an "uncle," periodically disappearing as the insults from family as well as possibly employer become increasingly intolerable.[10] Many aspects of these despised and pathetic men have been internalized in childhood. Now, as the black adolescent consciously acknowledges to himself his hatred of this father or uncle, his sensed similarities to the figure becomes highly problematic.[11] The presence of such despised internalized figures is still another intra-psychic condition giving rise to self-hatred.

Social Character: Images and Styles

The individual's self-hatred is intensified and perpetuated in large mea-sure by the process of "victimization."[12] As an adaptive response to his social conditions, the lower socioeconomic class Negro maintains himself as "victim":

"[whites] by their greater power, create situations in which Negroes do the dirty work of caste victimization for them.

"[to] Structural Conditions Highly Inimical to Basic Social Adaptation (low income availability, poor education, poor services, stigmatization) . . . Negroes adapt by personal and social responses which serve to sustain the individual in his punishing world but also generate aggressiveness towards the self and others which results in *suffering directly inflicted by Negroes upon themselves and others.*"[13]

Through such sustaining responses the Negro man contributes to per-petuating the *social character images* portraying him as debased and useless. One expression of these linked themes—low self-esteem and self-hatred—is in the Sambo image. Eloquently, and at length, depicted by Elkins, this configuration is one that resembles the "Uncle Tom" so disdainfully spoken of, yet so tenaciously held and acted upon:

"Sambo . . . was docile but irresponsible, loyal but lazy, humble but chronically given to lying and stealing; his behavior was full of infantile

[10] Liebow, particularly in the chapter "Fathers without Children," gives numerous examples of this; see *Tally's Corner*, 1968, pp. 72–102.

[11] The similarities to the hated father are strikingly shown in the high correlations between the *me* self-image and the *pa* self-image for the Negro boys.

[12] See Rainwater (1966), p. 175.

[13] *Ibid.*, p. 175; italics added.

silliness and his talk inflated with childish exaggeration. His relationship with his master was one of utter dependence and child-like attachment: it was indeed this childlike quality that was the very key to his being.[14]

"He [the white] might conceivably have to expect in this child [the Negro man]—besides his loyalty, docility, humility, cheerfulness, and (under supervision) his diligence—such additional qualities as irresponsibility, playfulness, . . . laziness, and (quite possibly) tendencies to lying and stealing. Should the entire prediction prove accurate, the result would be something resembling 'Sambo.' "[15]

Early, and it seems lasting, features of the Sambo image are quoted by Elkins in further painting the details of this pervasive character type:

"The Negro was to be a child forever. 'The Negro is always a boy, let him be ever so old. . . .' 'He is a dependent upon the white race; dependent for guidance and direction. . . . Apart from this protection he has the helplessness of a child—without foresight, without faculty of contrivance, without thrift of any kind.' Not only was he a child, he was a happy child. . . . Few Southern writers failed to describe with obvious fondness . . . the perpetual good humor that seemed to mark the Negro character, the good humor of an everlasting childhood."[16]

Elkins views Sambo as a "social role," the behavior expected of persons who belonged to the "specific social group" of Negroes. In analyzing the system of southern slavery much of his argument rests upon the existence of this role, *and* the playing of it by the southern Negro. However, as we have already seen, this social role—or *social character image*—remains a viable one at this time as well. Its outlines have been sketched many times over by the Negro subjects and by the numerous observers already cited.

It is important here to distinguish between the notion of social role, and *performance* of it, between expectations and behavior. To clarify the significance of the social character images of Sambo and Victim, performances must now be considered. One valuable way to consider performance is in terms of *character styles*,[17] consistent individual patterns of interpersonal relations and self-expectations.

There are two character styles highly relevant to Sambo and Victim. The first of these patterns, other-direction, was described by Riesman in

---

[14] Elkins. S., *Slavery*, Grosset and Dunlap. New York, 1963, p. 82.

[15] Elkins (1959), p. 131.

[16] Elkins (1959), p. 132. The first quotation is from E. Pollard, *Black Diamonds Gathered in the Darkey Homes of the South*, New York: P and Russell, 1859, p. viii; the second quotation is from J. Kennedy, *Swallow Barn*, Philadelphia: Carey and Lee, 1832.

[17] Another term for such patterns is "life styles"; see Rainwater (1966), p. 206.

The Lonely Crowd.[18] The style is one of pleasing others, "fitting in", at almost any cost. In discussing more privileged youth, Keniston discusses other-direction:

"A few young men and women attempt to find an alternative to identity in other-direction. Unable to discover or create any solid internal basis for their lives, they become hyperadaptable; they develop extraordinary sensitivity to the wishes and expectations of others. In a very real sense they let themselves be defined by the demands of their environment . . . they are safe from disappointments . . . he has settled for playing the roles others demand of him."[19]

The second character style—the expressive life style—is closely related to other-direction. The emphasis is again on the performance of roles according to others' expectations. However, there is an additional component in the pattern, that of "immediate gratification." The style is "an effort to make yourself interesting and attractive to others so that you are better able to manipulate their behavior along lines that will provide some immediate gratification."[20]

Both these styles provide the means for perpetuating the Victim and Sambo character images. To be accepted within the family, within the local community, the posture of Victim or at best "Unattractive" is essential for the black adolescent. His family is the first teacher of these postures:

"In most societies, as children grow and are formed by their elders into suitable members of the society they gain increasingly a sense of competence and ability to master the behavioral environment their particular world presents. But in Negro slum culture growing up involves an ever increasing appreciation of one's shortcomings, of the impossibility of finding a self-sufficient and gratifying way of living. It is in the family first and most devastatingly that one learns these lessons."[21]

Once learned, the images are maintained by the individual through continuous vigilance to the cues of those around him, particularly his family and neighborhood. The irony of all this is that through acting irresponsible, weak, and degraded a man makes himself more attractive to those around him. In so doing he ensures himself of a "safer" position:

"The man learns that he exposes himself to the least risk of failure when

[18] D. Riesman, *The Lonely Crowd*, 1950.
[19] K. Keniston, "Social Change and Youth in America," *Daedalus*, **91**, 163, 1962.
[20] Rainwater (1966), p. 206.
[21] *Ibid.*, p. 203.

he does not assume a husband's and father's responsibilities, but instead counts on his ability to court women and ingratiate himself to them."[22]

Rainwater observes that being Victim or Sambo is a requirement of the special set of conditions faced by the lower socioeconomic class Negro:

"The ghetto Negro is constantly confronted by the immediate necessity to suffer in order to get what he wants of those few things he can have . . . he suffers as exploited student and employee, as drug user, as loser in the competitive game of his peer group society. . ."[23]

By maintaining these devastating images of himself, the black man emerges with a "peculiar strength and a pervasive weakness":

"The strength involves the ability to tolerate and defend against degrading verbal and physical aggressions from others and not to give up completely. The weakness involves the inability to embark hopefully on any course of action that might make things better, particularly action which involves cooperating and trusting attitudes toward others."[24]

He is armed against others and now equipped to be accepted by a broad sector of persons. For by continuing to play these important roles, by continuing to live as an inferior, weak man with limited abilities and "potentials," he fulfills the general expectations of those outside as well as within his lower-class culture.[25]

There is a third style which is related to both these character images and the issues of self-limitation or identity foreclosure. This character style is generated by the same matrix of social conditions:

". . . increasingly as members of the Negro slum culture grow older, there is the *depressive strategy* in which goals are increasingly constricted to the bare necessities for survival. . . . This is the strategy of 'I don't bother anyone and I hope nobody's gonna bother me; I'm simply going through the motions to keep my body (but not soul) together.' "[26]

This last character style leads most explicitly to constriction of wishes and goals, the self-limitation so basic to identity foreclosure. The stasis is inherent in "the deadness of the depressed style."[27]

[22] *Ibid.*, p. 199.

[23] *Ibid.*, p. 176.

[24] *Ibid.*, p. 204.

[25] Pertinent to this point of fulfilling prophesies and "vicious circles," see Wender (1968).

[26] Rainwater (1966), p. 207. Rainwater uses the terms "strategy" and "style" interchangeably, seeing a particular style as a form of strategy.

[27] *Ibid.*, p. 207.

Yet this entire set of intertwined character images and character styles is in the direction of constriction. The images of Sambo and Victim, with all their implications, present well-defined, beautifully prepackaged character types.[28] The styles that emphasize *hyperadaptability* perpetuate the acceptance of these finely wrought roles, roles that have been rehearsed, played, and given standing ovations since at least the time of nineteenth century American slavery. Given such pervasive and meticulously constructed character images, along with the forces and mechanisms for their continuity, it becomes clear how the experience of *limitation* and *constriction* is generated for the individual black adolescent.

It is most important to remember that there are *external* restrictions on options for the Negro adolescents, together with the self-imposed limitations. These external aspects were sketched earlier. To an extent it could be argued these limitations are surmountable. Who surmounts them, and how, seems to be an important and unanswered question, worthy of careful study. In part the answer to this question probably is in terms of those privileged few who have not adopted the pervasive and in many ways "deadening" character styles and associated character images.[29]

Before considering a second explanatory model, the preceding framework may be reviewed by way of Figures 1 and 2. In these figures, through outline and diagram, the psychosocial matrix of identity foreclosure is summarized.

THE DEVELOPMENT OF IDENTITY FORECLOSURE

Thus far, we have analyzed the black identity foreclosure predominantly from the perspective of its presence as the end result of specific adaptations to sociocultural conditions. Although occasionally alluded to, the problem of historical roots has not yet been fully explored in the discussion. In other words, we have only inquired how identity foreclosure is *perpetuated*, by both the individual and his surrounding sociocultural milieu.

A second critical question must also be asked, namely, what leads to the emergence of identity foreclosure in the individual life cycle? What

[28] In discussing vocational choices of college students, Beardslee and O'Dowd comment on the early choice of occupation as related to the stereotyped styles of life perceived for that role. In such situations, for probably very different underlying reasons, a state of identity foreclosure is also present. See D. Beardslee and D. O'Dowd, "Students and the Occupational World," in N. Sanford (Ed.), *The American College*, New York, Wiley, 1962.

[29] Available data for analysis of such questions is probably in autobiographical works as *Manchild in the Promised Land* (1965), *The Autobiography of Malcolm X* (1967), and *Soul On Ice* (1968).

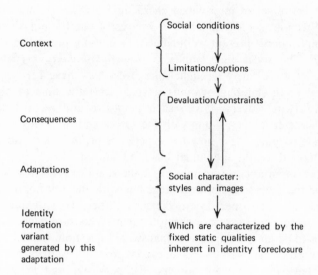

Figure 1.   The context that evolves and maintains identity foreclosure: a summary.

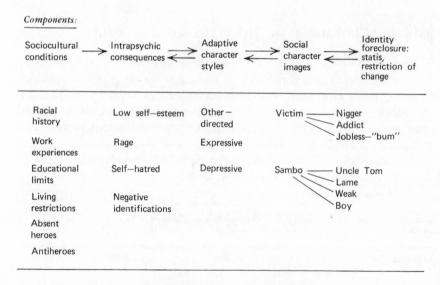

Figure 2.   Intrapsychic consequences and adaptations.

conditions are necessary for this identity variant to occur for the lower socioeconomic class black adolescent?

The notion of identity foreclosure appears in Erikson's discussions of precursors of the adolescent identity crisis.[30] His epigenetic model of personality development assumes a "progression through time of a differentiation of parts." This means:

"(1) that each item of the healthy personality . . . *is systematically related to all others, and that they all depend on the proper development in the proper sequence of each item;* and (2) that each item *exists in some form before 'its' decisive and critical time normally arrives.*"[31]

The preadolescent forms of ego identity are present in the latency period. It is within this stage that "competence" is a prominent concern for the developing child:

". . . he must begin to be somewhat of a worker and potential provider before becoming a biological parent. With the oncoming latency period, the normally advanced child forgets, or rather "sublimates" (that is, applies to more useful pursuits and approved goals) the necessity of "making" people by direct attack or the desire to become mother or father in a hurry: he now learns to win recognition by *producing things.* He develops industry . . . to become an eager and self-absorbed unit of a productive situation to completion is an aim which gradually supercedes the whims and wishes of his idiosyncratic drives and personal disappointments. . . . He now wants to make things well. . . . He develops the pleasure of work completion by steady attention and perservering diligence."[32]

Two general conditions are required for the development of this competence: previously successful steps in earlier ego development and a facilitating environment within the latency period. If these requirements are fulfilled, the set of conflicts—or "crisis"—of latency then resolves and the developing child emerges with "a sense of industry." The identity precursors of such a successful resolution are positive "work identifications."[33]

[30] E. Erikson, "The Problem of Ego Identity," *Psychological Issues,* **I**(1), 120, 1959.

[31] E. Erikson, "Growth and Crises of the 'Healthy' Personality," *Psychological Issues,* I(1), 53, 1959; italics are Erikson's.

[32] E. Erikson, "Growth and Crises of the 'Healthy' Personality," p. 86.

[33] E. Erikson, "Ego Identity and the Psychological Moratorium," in H. Witmer (Ed.), *New Perspectives for Research on Juvenile Delinquency,* Washington, D.C.: U. S. Government Printing Office, 1955, p. 8.

The successful resolution is represented in the next stage of ego development—adolescence—in the sense of "anticipation of achievement."[34]

However, if the individual fails to resolve these latency issues, he emerges from this period with a sense of "inadequacy and inferiority." The identity precursor corresponding to this developmental failure is *identity foreclosure,* a premature interruption in the adolescent task of identity formation. There are two seemingly opposite forms in which this precursor can be expressed. It can appear *negatively,* as the "sense of inferiority, the feeling that one will *never* be any good,"[35] or it can appear *positively,* when, for example:

"[the child] identifying too strenuously with a too virtuous teacher or becoming the teacher's pet . . . his sense of identity can become prematurely *fixed* on being nothing but a good little worker or a good little helper, which may not be all he *could* be."[36]

The predominant form of identity foreclosure for the black subjects was the negative one. Repeatedly in the interviews, as illustrated earlier, there emerged the themes of inferiority, mediocrity, and degradation. Sufficient data is not available to firmly reconstruct the latency of these subjects. Their accounts of work situations as adolescents certainly encourages the speculation that experiences of failure at work, be it with school or "jobs," occurred earlier in their lives as well. The black youth complained with great frequency of "being looked down upon" in school and of being given second-rate work opportunities, "where doors are slammed at you." Events such as these in latency would be compelling conditions for the generation of a sense of inferiority and the corresponding identity precursor, identity foreclosure.

There are intrapsychic as well as other sociocultural determinants that may underlie the development of a "sense of inadequacy and inferiority" with its concomitant identity foreclosure:

"[The sense of inadequacy] may be caused by an insufficient solution of the preceding conflict: he may still want his mummy more than knowledge; he may still rather be the baby at home than the big child in school; he still compares himself with his father and the comparison arouses a sense of guilt as well as a sense of anatomical inferiority. Family life (small

---

[34] *Ibid.*

[35] E. Erikson, "Growth and Crises of 'Healthy' Personality," p. 87; italics added. It is of interest that Erikson adds that this is "a problem which calls for the type of teacher who knows how to emphasize what a child *can* do and who knows a psychiatric problem when she sees one."

[36] *Ibid.*, p. 88; italics of "fixed" are added.

family) may not have prepared him for school life, or school life may fail to sustain the promises of earlier stages in that nothing he has learned to do well already seems to count one bit with the teacher. And then again, he may be potentially able to excell in ways which are dormant and which, if not evoked now, may develop late or never."[37]

One type of identity foreclosure is a derivative of the stage of development preceding latency. This foreclosure is one of *total negation*, of repudiation. It is based on "all those identifications and roles which, at critical stages of development, had been presented to the individual as most undesirable or dangerous. . . ."[38] This foreclosure is that of *negative identity*.[39] The individual's identity configuration is prematurely fixed on the repudiated, the personally scorned and rejected identifications and roles.

Negative identity is the adolescent derivative of a failure to resolve the set of issues belonging to the developmental stage of conscience formation. It is this phase of ego development in which the conflicts between guilt and initiative become most prominent. Where guilt predominates, there results "a *self-restriction* which keeps an individual from living up to his inner capacities or to the powers of his imagination and feeling."[40] This pattern of self-restriction is already familiar as a fundamental aspect of identity foreclosure. However, where conflicts over initiative and guilt are of great magnitude, there may emerge *paralysis* of ambition and a choice of the despised as the focus of one's self-image. This configuration, then, is negative identity:

"Such vindictive choices of a negative identity represent of course a desperate attempt at regaining some mastery in a situation in which the available positive identity elements cancel each other out. The history of such a choice reveals a set of conditions in which it is easier for the patient to derive a sense of identity out of a total identification with that which he is least supposed to be than to struggle for a feeling of reality in acceptable roles which are unobtainable with his inner means . . . the relief following the total choice of a negative identity."[41]

One expression of negative identity, of relevance here, is that of "snob-

---

[37] *Ibid.*, p. 86.

[38] E. Erikson, "The Problem of Ego Identity," p. 131.

[39] In Erikson's recent writings there is interesting support for viewing negative identity as a subtype of identity foreclosure. Instead of using the term "negative identity" in speaking of the derivative for adolescence of a developmental failure in the oedipal period, now the term *"role fixation"* is employed. See "The Problem of Ego Identity," p. 144 and *Identity: Youth and Crisis*, p. 94.

[40] Erikson (1968), p. 120; italics added.

[41] *Ibid.*, p. 176.

bism." Erikson discusses forms of snobbism as a response to the danger of identity diffusion. Confronted with the adolescent conflicts of identity formation and the possibility of failing to resolve them, an earlier developmental faliure becomes most prominent:

". . . they [upper-class forms of snobbism] permit some people to deny their identity confusion through recourse to something they did not earn themselves, such as their parents' wealth, background, or fame or to some things they did not create. . . . But there is a *"lower-lower"* snobbism too, which is based on a pride of having achieved a semblance of nothingness."[42]

Negative identity is a highly specific type of identity foreclosure. Identity formation is prematurely halted through a commitment of the individual to that which is alien, personally most shunned. In making so *total* a choice, the ambiguity and conflicts inherent in multiple identifications are eliminated in favor of a single principle, a single set of identifications. The identity rests on the despised. The individual considers himself as hateful, *totally* undesirable.

In another context Erikson refers to this identity pattern as a "total commitment to role fixation."[43] Although this fixation may become most obvious in adolescence, its genesis is in the oedipal period of life. Failure to deal with and, on balance, to overcome the conflicts over guilt and initiative have facilitated the emergence of a negative identity in adolescence:

". . . the display [negative identity] . . . has an obvious connection with earlier conflicts between free initiative and oedipal guilt in infantile reality, fantasy, and play . . . the choice of a *self-defeating* role often remains the only acceptable form of initiative . . . this in the form of a complete denial of ambition as the only possible way of totally avoiding guilt."[44]

Negative identity represents an identity configuration that is an integration of hated and repudiated identifications. It is thus a configuration of self-repudiation, or self-negation. Along a continuum of identity foreclosures, negative identity represents the negative pole. Between this polar type and the positive polar type—the "precocious genius" or "pet"—lie the many other variations of identity foreclosure. Within these poles are varieties of "inferiority" and "superiority." The clinical data suggest that the black subjects' identity foreclosure falls in the negative range of such a continuum. They consider themselves "inferior," limited, "lame," and therefore fit *only* for a certain niche.

[42] *Ibid.*, p. 176; italics added.
[43] *Ibid.*, p. 184.
[44] *Ibid.*

Yet within this group of adolescents the extreme polar types, negative identities, may also exist. Careful scrutiny of early personal history, early identifications, and of the content of a subject's multiple self-images *as well as* their structural relationships would be required to discover the presence of negative identity configurations. Are the current self-images of a subject, for instance, consistent with those roles and figures he was taught were most undesirable? Do the subject's highly valued self-image features correspond to those features of others that were regarded in early years as despicable and degraded?

Disturbances in two different developmental periods may underly identity foreclosure patterns. Clearly, the roots of these disturbances are multiple. For the black subjects it is likely that issues of work in latency play a prominent role in the identity foreclosure. However, a number of other conditions have also suggested themselves as significant in the genesis of this most striking variant, among them being the important figures, the fathers, the heroes, and the antiheroes. In considering negative identity as a type of foreclosure, the whole area of figures becomes extremely important as a realm of determinants for these patterns.

In some respects, the two models presented have overlaps. Following a diagrammatic presentation of this second model (Figure 3), relationships between the two models and remaining questions are reviewed.

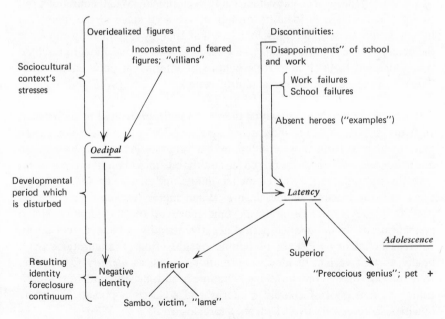

Figure 3. Developmental model of identity foreclosure.

THE MODELS

Both models, it is recalled, emphasize the sociocultural conditions—the contexts—within which identity foreclosure emerges. There are, it seems, at least two general patterns of relationship between the context, or complex "reality," and identity foreclosure.

The first pattern is that of *adaptation* and maintenance. The pattern can be described by way of several questions. Given a set of sociocultural conditions, what are the enduring identity configurations of the individuals confronting these conditions? What are recurrent behavioral and emotional styles that appear in such a context? And finally, perhaps most significantly, what are the dynamics of this "enduring" aspect; in other words, *how* are the configurations maintained? The relationship pattern is an equilibrium established between the individual and his complex sociocultural setting. Analysis of the components of this equilibrium and their interactions with one another is the purpose of the first of the models presented above.

The second relationship pattern is a more *historically oriented* one. For a particular identity configuration to be maintained, it obviously must have been present from the start. This pattern deals with the impacts of a sociocultural setting on individual ego development. What conditions led to a particular form of identity formation? At what stage of the individual life cycle did these conditions intersect and thereby significantly influence the emergence of a particular identity variant? The second model presented was addressed to analysis of this pattern of interaction between the individual's life stage and his sociocultural context.

There are several ways in which the two models are related to each other. In figure 4 such links are indicated diagrammatically. To begin with, both patterns emerge from those conditions that generate individual devaluation and constraint. When encountered in adulthood, these conditions are dealt with through adaptations of character image and character style, adaptations with the features of foreclosure. When met in earlier developmental periods, however, the result is the emergence of specific identity-related configurations. In the oedipal period negative identity is the eventual result, whereas in latency a "sense of failure" results. Both configurations lead, finally, to that variant of identity formation known as *identity foreclosure*. In addition, these configurations—"failure" and negative identity—contribute to that pool of character images and styles that possess the static qualities so representative of identity foreclosure.

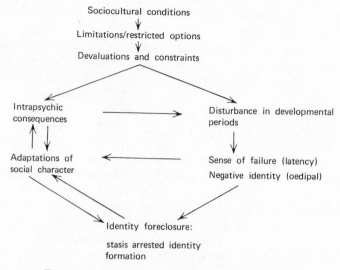

Sociocultural conditions

Limitations/restricted options

Devaluations and constraints

Intrapsychic consequences

Disturbance in developmental periods

Adaptations of social character

Sense of failure (latency)
Negative identity (oedipal)

Identity foreclosure:

stasis arrested identity formation

Figure 4. Comparison of the two models.

QUESTIONS AND DILEMMAS

Remaining, to be sure, are a multitude of unanswered and fascinating questions. Many have been raised in the previous chapters, others have been but faintly alluded to. The most obvious, and perhaps most pressing, question is that of *generality*. How characteristic of all lower socioeconomic class Negroes is identity foreclosure? Is this result an artifact of the small sample, no matter how statistically significant it appears? Or, and more likely, is this variant a predominant one which will be found in larger populations of lower socioeconomic class Negro youth as well? Is the variant also a prevalent one among a similar age and class group of Negro females? Do middle class or "upwardly mobile" lower class, Negro men also have this variant in identity formation?

A second set of questions raised concerns relationships within the study data itself. The interviews suggested trends of "hero deprivation" among the Negroes. Are such trends in some way correlated with the identity variant that was also found within this group? If they are, what then are the ways in which these relationships are mediated? Are there rigorous means by which one can verify these suggested hero trends?

To further develop as well as answer these two important sets of questions, a new study has begun with larger samples of black and white adolescents. In addition to the Q-sort, also available now is a design for studying heroes. But, there are other, perhaps even more difficult, dilemmas which must be recognized and for the time being go unanswered although respected as serious ones:

1. Are there means for carefully studying identity formation in terms of thematic aspects? Can one construct definitions of identity variants on the basis of prominent issues and themes as well as through formal—structural —qualities?

2. Identity formation in the nonadolescent stages of the life cycle has been largely unexplored here or elsewhere. With the existence of an instrument such as the Q-sort, other age groups become highly amenable to controlled study and comparison of identity formation. There are obviously fascinating questions in this area. What, for instance, is the nature of identity development at different ages? What formal characteristics are found in childhood, latency, and old age?

3. The original question about the universality of identity formation is still a meaningful one. How does identity formation vary in other sociocultural groups within the United States? Can one find known variants of identity formation in other cultural groups or is this developmental pattern even found in these other groups? Again, the presence of the Q-sort encourages pursuit of these intriguing cross-cultural developmental questions.

4. What further theoretical and empirical considerations must be made to explain the identity variants found in these black and white populations? How can the models be further refined? Or are there other models even more relevant to this data?[45]

5. Finally, perhaps of greatest interest among these many issues, have the apparently rapid changes in black-white relations influenced identity formation patterns? At this time, five years after the original study, would one no longer find predominant foreclosure patterns among lower-class black youths?

In addition to theoretical and empirical concerns, there are social concerns generated by these findings. One cannot lose sight of the discouraging implications of identity foreclosure. This variant reflects an arrest in ego development, a halt in the growth from child to man. If this variant is so prevalent among lower-class black male youth as the findings suggest, this means that a population of individuals has in some way been crippled in its development. It then becomes crucial to search for those conditions that underlie this growth failure, this deterrent to the fulfillment of potentials. Indeed, to attend to systematic and rigorous study of this variant and its roots represents more than theoretical and empirical curiosities.

[45] A new study, currently in progress, is investigating cognitive aspects of adolescent development in these two groups. Thus other theoretical perspectives will be introduced.

To obtain insight into the environmental and psychological conditions surrounding foreclosure is of immense importance. The consequences of this variant in identity formation are clear: wasted potentials and abortive growth. Any steps to thwart such trends in development require application of insights into the trends themselves and careful understanding of what generates as well as sustains patterns of identity foreclosure.

Epilogue

The subject of values and "neutrality" is intimately related to this investigation and this monograph. The issue of whether or not social science research can—or should—be wholly free of the researcher's values has been discussed for many years. It would not be useful at this point to review the controversy. However, I do want to make my position clear, as it applies to this study. As much as possible, the repeated objective testing of the subjects was performed without apparent bias or preconception. However, the nature of the results and of their implications brings forth explicit value judgements.

In the final chapters several interpretations of the results were offered. Yet, implied throughout the discussion is a more absolute position as well. There are aspects of these results that reach beyond merely different interpretations.

These findings are consistent with an interruption in adolescent development for the black adolescents. If we use as absolute standards of health those of fullest development, then it is clear that the blacks' chances for such health were substantially reduced. Alternatively, one might argue that the black identity patterns represent different "life styles," differing adaptations to oppressive living conditions. I think the argument must then turn to the fact that some of these styles are in the long run hardly adaptive, for they are manifestations of abortive development. The results indicate diminished psychological growth for the black adolescents, a probable consequence of many decades of oppression. They are disturbing results and they urge that we examine as fully as possible the social conditions that may have contributed to such adolescent growth failures. In addition, the variant lines of development associated with these conditions must also continue to be most rigorously investigated. It is as we gather this kind of understanding that we can more effectively assist in maximizing rather than inhibiting the growth of the individual.

APPENDIX A

Self-Image Significance Tables

For the purpose of complete recording of all comparative correlation data, several cutting points are used in reporting the following results of nonparametric tests:

1. All differences at or below .05 probability are given at their exact probability levels. These differences are clearly *significant* ones.

2. All differences which are at probability levels between .05 and .20 are also reported at their exact probability levels. These differences represent *tendencies* in the findings.

3. All differences which are at a probability level above .20 are reported as >.20. These results suggest, at most, *trends* in the data, and are not treated as interpretable findings in any more specific manner than this.

4. Where no differences at any probability level were found, the result is reported as "B = W."

In interpreting the data[1] it is obviously the results occurring at or below .05 probability which are most important and most relied upon. However, for the sake of completeness and appreciation of tendencies and trend patterns, the other two cutting points are also included.

---

[1] The complete matrices of correlations from all subjects can be obtained by contacting the author.

Table 1. Interracial Comparisons of Parental Self-Image: Intrayear
Pearson Correlations

		me/pa		me/ma		me/ma	
Year	N^a	Comparison	P^b	Comparison	P	Comparison	P
1	10 B; 10 W	B > W	.05	B > W	.10	B = W	—
2	7 B; 6 W	B > W	.04	B > W	.03	B > W	.10
3	7 B; 5 W	B > W	.10	B > W	.13	B > W	> .20
4	7 B; 4 W	B > W	.13	B > W	> .20	W > B	.14

[a] The symbol "B" represents black subjects; the symbol "W" represents
white subjects.
[b] As determined by the Mann-Whitney U Test.

Table 2. Interracial Comparisons of Parental Self-Images:
Interyear Pearson Correlations

Year Interval	N	pa		ma	
		Comparison	P^a	Comparison	P
1–2	7 B; 6 W	B > W	> .20	B > W	.18
1–3	7 B; 6 W	B > W	> .20	B > W	> .20
1–4	7 B; 4 W	B > W	.16	B > W	> .20
2–3	5 B; 5 W	B > W	.13	B > W	.11
2–4	5 B; 4 W	B > W	.03	B > W	.11
3–4	7 B; 4 W	W > B	> .20	B > W	> .20

[a] As determined by the Mann-Whitney U Test.

Table 3. Interracial Comparisons of Personal Time: Self-Image Intrayear Pearson Correlations

		me/past		me/future		past/future		future average	
Year	N	Comparison	P[a]	Comparison	P	Comparison	P	Comparison	P
1	10 B; 10 W	B = W	—	B = W	—	B = W	—	B > W	.05
2	7 B; 6 W	B > W	> .20	B > W	.13	W > B	.20	B = W	—
3	7 B; 5 W	B = W	—	B > W	.17	B = W	—	B > W	> .20
4	7 B; 4 W	B = W	—	B = W	—	W > B	.20	B > W	> .20

[a] As determined by the Mann-Whitney U Test.

Table 4. *Changes* of Black Subjects: Intrayear Pearson Correlations for Personal Time Self-Images

		me/past		me/future		past/future		past average	
Year Interval	N	Comparison[b]	P[a]	Comparison	P	Comparison	P	Comparison	P
1–2	7 B; 6 W	Same	—	Same	—	Lower	.20	Same	—
1–3	7 B; 6 W	Same	—	Higher	.05	Same	—	Same	—
1–4	7 B; 4 W	Lower	.20	Higher	.05	Same	—	Same	—
2–3	5 B; 5 W	Same	—	Same	—	Higher	.20	Lower	.20
2–4	5 B; 4 W	Lower	.20	Higher	.06	Lower	.20	Lower	.20
3–4	7 B; 4 W	Lower	.20	Same	—	Lower	.20	Lower	.20

[a] As determined by Wilcoxon and sign tests.
[b] Entry of "same," "lower," or "higher" refers to comparison of later year with earlier year.

Table 5. *Changes of White Subjects' Intrayear Pearson Correlations for Personal Time Self-Images*

Year Interval	me/past		me/future		past/future		past average	
	Comparison	P^a	Comparison	P	Comparison	P	Comparison	P
1-2 7 B; 6 W	Same[b]	—	Same	—	Same	—	Same	—
1-3 7 B; 6 W	Lower	>.20	Higher	>.20	Lower	.06	Lower	>.20
1-4 7 B; 4 W	Same	—	Higher	4/4[c]	Same	—	Lower	4/4
2-3 5 B; 5 W	Same	—	Same	—	Lower	>.20	Lower	>.20
2-4 5 B; 4 W	Same	—	Higher	4/4	Same	—	Lower	4/4
3-4 7 B; 4 W	Same	—	Higher	4/4	Same	—	Same	—

[a] As determined by Wilcoxon and sign tests.

[b] Entry of "same," "lower," or "higher" refers to comparison of later year with earlier year.

[c] Result reported as such because specific probability level unavailable for this size N.

Table 6. Interracial Comparisons of Personal Time Self-Images:
Interyear Pearson Correlations

Year Interval	N	past		future	
		Comparison	P^a	Comparison	P
1-2	7 B; 6 W	B > W	.08	B > W	.08
1-3	7 B; 5 W	B > W	> .20	B > W	> .20
1-4	7 B; 4 W	B > W	> .20	B > W	.13
2-3	5 B; 5 W	B > W	> .20	B > W	.13
2-4	5 B; 4 W	B > W	> .20	B > W	> .20
3-4	7 B; 4 W	W > B	> .20	B = W	—

[a] As determined by the Mann-Whitney U Test.

Table 7. Interracial Comparisons of *fantasy* Self-Image:
Interyear Pearson Correlations

Year Interval	N	fantasy	
		Comparison	P^a
1-2	7 B; 6 W	B > W	.10
1-3	7 B; 6 W	B = W	—
1-4	7 B; 4 W	B > W	.16
2-3	5 B; 5 W	B > W	> .20
2-4	5 B; 4 W	B > W	.04
1-4	7 B; 4 W	B > W	.16

[a] As determined by the Mann-Whitney U Test.

Table 8. Interracial Comparisons of Current Self-Image (*me*):
Average Intrayear Pearson Correlations

Year	N	Average me^a	
		Comparison	P^b
1	10 B; 10 W	B = W	—
2	7 B; 6 W	B > W	.16
3	7 B; 5 W	B > W	> .20
4	7 B; 4 W	B = W	—

[a] The average *me* is calculated using *z* conversions of the seven other self-images with which it occurs in each of the years; for example, *me/ma*, *me/friend*, and so on.

[b] As determined by the Mann-Whitney U Test.

Table 9. Interracial Comparisons of Current Self-Image (*me*):
Interyear Pearson Correlations

Year Interval	N	me Comparison	P[a]
1–2	7 B; 6 W	B = W	—
1–3	7 B; 6 W	B > W	> .20
1–4	7 B; 4 W	B > W	> .20
2–3	5 B; 5 W	B > W	.03
2–4	5 B; 4 W	B > W	.12
3–4	7 B; 4 W	B > W	.10

[a] As determined by the Mann-Whitney U Test.

Table 10. Interracial Comparisons of Individal Subject's *Changes* in Averages
of Intrayear Pearson Correlations (*Structural Integration*)[a]

Year Interval	N	Black	P[b]	White	P
1–2	7 B; 6 W	Same	—	Same	—
1–3	7 B; 6 W	Higher	.12	Lower	> .20
1–4	7 B; 4 W	Same	—	Higher	3/4[c]
2–3	5 B; 5 W	Higher	> .20	Same	—
2–4	5 B; 4 W	Same	—	Higher	3/4
3–4	7 B; 4 W	Same	—	Higher	3/4

[a] This table shows the blacks with tendencies towards increasing values in intervals Years 1–3 and 2–3. Such a difference does not appear in Figure 16. The apparent discrepancy it attributable to the fact that the table is calculated from *matched-pair* comparisons (same subject), whereas the figure represents group tendencies (medians of the entire racial group) and thus even includes some values not included in the table. Despite this difference in the two displays of data, the table also again demonstrates that the whites are the subjects who are the "changers."

[b] As determined by the sign test using comparisons of averages for same subjects.

[c] Where fractions are given, they represent the actual number who change in the stated direction, as compared to the total number of subjects in that particular interval.

Table 11. Interracial Comparisons of Interyear Differences in Yearly Averages of Intrayear Pearson Correlations (*Structural Integration*)

Year Interval	N	Black Intrayear Averages	White Intrayear Averages	P^a
1–2	7 B; 6 W	$B_2 = B_1{}^b$	$W_2 = W_1$	—
1–3	7 B; 6 W	$B_3 = B_1$	$W_3 > W_1$	.06
1–4	7 B; 4 W	$B_4 = B_1$	$W_4 > W_1$	.20
2–3	5 B; 5 W	$B_3 = B_2$	$W_3 = W_2$	—
2–4	5 B; 4 W	$B_4 = B_2$	$W_4 > W_2$	$> .20$
3–4	7 B; 4 W	$B_4 = B_3$	$W_4 = W_3$	—

[a] As determined by the Mann-Whitney U Test. Although the years have, in part, the same Ss, several years have different numbers of Ss. To use a related samples test would therefore have meant excluding several results. Hence the Mann-Whitney test is again applied here.

[b] Numerical subscript refers to the corresponding year.

Table 12. Interracial Comparisons of Average Interyear Pearson Correlations (*Temporal Stability*)

Year Interval	N	Interyear Averages	P^a
1–2	7 B; 6 W	B > W	.04
1–3	7 B; 6 W	B > W	.12
1–4	7 B; 4 W	B > W	.20
2–3	5 B; 5 W	B > W	.20
2–4	5 B; 4 W	B > W	.04
3–4	7 B; 4 W	B = W	—

[a] As determined by the Mann-Whitney U Test.

Table 13. Comparisons of Blacks with Fathers versus Blacks without Fathers for the Magnitude of Their *pa* Interyear Correlation

		pa	
Year Interval	N	Comparison	P[a]
1–2	7	Bo > Bp[b]	.15
1–3	7	Bo > Bp	.03
1–4	7	Bo > Bp	.20
2–3	7	Bo > Bp	.10
2–4	7	Bo > Bp	.20
3–4	7	Bo > Bp	.20

[a] As determined by the Mann-Whitney U Test.
[b] Bo = subjects without fathers; Bp = subjects with fathers.

Table 14. Comparisons of Blacks with Fathers versus Blacks without Fathers for Magnitude of Their *me/pa* Correlation

		me/pa	
Year	N	Comparison	P[a]
1	10	Bo > Bp[b]	.09
2	7	Bo > Bp	.06
3	7	Bo > Bp	.06
4	7	Bo = Bp	—

[a] As determined by the Mann-Whitney U Test.
[b] Bo = subjects without fathers; Bp = subjects with fathers.

APPENDIX B

Further Aspects of the Sample

Table 1. Additional Characteristics of Participants and Dropouts[a]

Subject	Race[b]	Geographic Mobility	School Performance[c]	School Dropout	Part-Time Work	Number of Siblings
Dropouts						
JE	B	—[d]	Average–above average	—	—[d]	2
KB	B	—	Average–above average	—	—	1
DM	W	—	Below average	+	+	3
AL	W	+	Average–above average	—	+	3
ND	W	+	Below average	—	+	3
IQ	W	—	Failing	+	+	8
LE	W	—	Below average	?	+	0
Active participants						
FM	B	+	Below average	—	+	5
GR	B	+	Average	—	—	5
LT	B	—	Below average	—	+	2
OS	B	—	Below average–failing	—	+	4
FL	B	—	Low average	+[e]	+	4
MD	B	—	Below average	—	+	1
TM	B	—	Average	—	+	4
BA	B	+	Low average	+	+	2
KE	B	—	Average	—	+	3
HN	PR	—	Average	—	+	5
NS	W	+	Low average	—	+	5
BT	W	—	Below average	—	+	6
VR	W	+	Average–above average	+[f]	+	1
JK	W	+	Average	—	+	2
JR	W	+	Average	—	+	2

[a] Other data, such as age and socioeconomic class are given in Table 1, Chapter 2.

[b] B, black; W, white; PR, Puerto Rican.

[c] As summarized from school records of the subject's last year of school attendance.

[d] —, Absence of a characteristic; +, presence of characteristic.

[e] Forced to leave school as a result of juvenile delinquency and subsequent reformatory sentence.

[f] Two-weeks duration.

APPENDIX C

Symbolic Formulations of Identity Variants

Listed in this section are "translations" of the operational definitions given in Chapter 3. The equations included here are abbreviations of complex processes. They are presented not to suggest that these processes can be simplified to one-line equations. The point, rather, is to suggest that there may be other forms of notation with which one can characterize these variants as empirical definitions and understandings of the relevant variables are further elaborated.

Preceding each symbolic formulation is a summary of the variant's definition.

1. *Progressive Identity Formation.* Throughout childhood and adolescence, an ego identity develops, with an acceleration of this process occurring in adolescence. Less specific remarks have been directed toward the period after adolescence. It is implied, however that the process slowly continues, reaching a climax in old age, the "wisdom" stage of the life cycle (Erikson, 1960). We may describe an individual as manifesting identity formation when both the structural integration and the temporal stability of his self-images are simultaneously increasing over any given period of time, though not necessarily at the same rate.

Symbolically: $S * T = F$, where S = structural integration and may only increase; T = temporal stability and may only increase; F = a dependent variable, which may only increase; $*$ = an interaction between S and T.

2. *Identity Diffusion.* This clinical type designates those cases in which there is failure to achieve integration and continuity of self-image. The category is a broad one and probably includes several subtypes, for there are conceivably multiple etiologies underlying this outcome of identity formation. Such a state may be present at any stage of the life cycle. However, it is theoretically most manifest at adolescence when it hinders psychosocial development. Again, only through further empirical investigation can this hypothesis of stage specificity be studied. For example, does identity diffusion occur during latency, or unexpectedly in adulthood, and lead to analogous complications in further psychosocial development?

Operationally, a person is said to be in a state of identity diffusion when both features of his self-images show a repeated decrease over any given period of time. Again, these changes need not be occurring at the same rate. In fact, differing rates may be one sign of variant kinds of identity diffusion. However, of the two features, structural integration is the most critical here in defining the state of diffusion. A third possibility in this category is a form of "attenuated diffusion," in which either (a) structural integration decreases and temporal stability remains constant; or (b) the milder type, in which temporal stability decreases and structural integration remains constant. It is possible that these represent early forms of flagrant identity diffusion, where reversibility is more likely.

Symbolically: $S * T = D$, where S and T are as previously defined and now may only decrease over time; D = a dependent variable which may only decrease over time; $*$ = an interaction between S and T.

Attenuated forms of diffusion are. Variant (a): $S * T = D$, where S must decrease over time; $T = K$ = constant; D = a dependent variable which decreases over time; $*$ = as defined above. Variant (b): $S * T = D$, where T must decrease over time; $S = K$ = constant; D = a dependent variable which must decrease over time; $*$ = as defined above.

3. *Identity Foreclosure.* This state superficially resembles identity development. There is a sense of integration, "purpose," stability, and a diminution in subjective confusion about these matters. However, the stability and purpose are reflections of an avoidance of alternatives, of a certain restrictiveness which eliminates any ambiguities. What appears to be the outcome of a successful process of identity formation is actually an impoverished, limited self-definition and sense of continuity. Operationally, a person is said to manifest identity foreclosure when either the structural integration and temporal stability of his self-images remain stable, or only temporal stability shows continued increase while structural integration is unchanging (See Chapter 3, footnote 7).

Symbolically: (a) $S = T = K$, where K is a constant; S and T are defined as above. (b) $S * T = C$, where C is a dependent variable which may only increase; K is a constant; S and T are defined as above; and $S = K;$ $*$ = interaction between S and T.

4. *Psychosocial Moratorium.* An individual is described as being in this state when he is "finding himself," experimenting with varied roles, new self-images, and future plans, at all costs remaining uncommitted to any particular alternative or identity. At the same time he is *not* tending toward, or in, a type of diffusion. This is, of course, the antithesis of foreclosure. Rather than rigid sameness, the content and types of self-images show continual variation. The key concept here is "openness," noncommitment. No irreversible decisions or plans are undertaken. A partly conscious, partly unconscious attempt is made to insure maximum flexibility and diversity before the further elimination of any possibilities, alternatives, or actions inherent in all stages of identity formation. Hypotheses as to when this stage occurs, and for whom, have been offered by Erikson.

Operationally, a person is in the period of psychosocial moratorium when the temporal stability of his self-images show significant fluctuations (increasing and decreasing) over a given period of time. Consistent with the tendency running counter to diffusion, the structural integration feature shows less fluctuation than that of temporal stability, particularly in terms of any decrease in value.

Symbolically: (a) When the two functions change but in opposite directions over time, thus "compensating" for each other: $S * T = M$, where S and T are as defined above; $M = K = $ constant; $* = $ as defined above. (b) $S * T = M$, where S is defined as above; $S = K$; $K = $ a constant; T is as defined above, and must alternately increase and decrease over time; $M = $ a dependent variable which increases and decreases over time. (c) $S * T = M$, where S is as defined above; T is as defined above; and both must independently increase or decrease over time; with any decrease in S being limited to *less than* $\frac{1}{2} \Delta T$. $M. = $ a dependent variable which increases and decreases over time.

APPENDIX D

Examples of Q-Sort Statements

Table 1. Black Subjects: *me* Self-image[a]

Rated high ("most important")

I have heart
I look at myself before I call someone a creep
I think we should all go back to Jamaica
I'm glad of my complexion
I like to travel
I build models
I want to be a good doctor
I want to be ambitious
I like someone who helps
I want to be an electrician
I want enough luck to get a job
I wish I were working now

Rated low ("least important")

I'm called the "mad archer"
I'm strong
I break windows
I think I'll buy a gas station
I have 15 girl friends
I think the biggest thing kids talk about is clothes
I always wanted to be a scientist
I'm the most popular guy
I like to wrestle
I like to do better

[a] Selected from three subjects over several years.

Table 2. White Subjects: *me* Self-image[a]

Rated high ("most important")

I score a lot in basketball
I want to go to prison rather than to Russia
I want to go into a good business
I know everything going on in my girl friend
I fight for freedom
I go along with styles
I'm too young to get serious
I like shorts
I want the Communist world wiped out

Rated low ("least important")

I like shoes with buckles on the side
I'm against my wife working
I know 400 coloreds
I fight for civil rights
I'm underweight
I like doctors to help me
I like to be like the Pope
I like to be like John Wayne

[a] Selected from two subjects over several years.

Bibliography

Ackerman, N. W. (1951). Social role and total personality. *American Journal of Orthopsychiatry*, **21**:1–17.

Adams, W. A. (1950). The negro patient in psychiatric treatment. *American Journal of Orthopsychiatry*, **20**:305–310.

Adelson, J. (1961). The adolescent personality. Paper read at the meeting of American Psychological Association, 1961, mimeo.

Aichorn, A. (1935). *Wayward Youth*. New York: Viking Press.

Antonovosky, A., and Lerner, M. (1959). Occupational aspirations of lower class negro and white youth. *Social Problems*, **7**:132–138.

Baittle, B. (1961). Psychiatric aspects of the development of a street corner group. *American Journal of Orthopsychiatry*, **31**:703–712.

Baldwin, J. (1953). *Go Tell It on the Mountain*. New York: Knopf.

Baldwin, J. (1961). *Nobody Knows My Name*. New York: Dial Press.

Baldwin, J. (1962). *Another Country*. New York: Dial Press.

Baldwin, J. (1963). *The Fire Next Time*. New York: Dial Press.

Barber, B. (1961). Social-class differences in educational life-chances. *Teachers College Record*, **63**:94–101.

Beardslee, D., and O'Dowd, D. (1962). Students and the occupational world. In Sanford, N. (Ed.), *The American College*. New York: Wiley pp. 597–626.

Bernard, J. (Ed.) (1961). Teen age culture. *Annals of the American Academy of Political and Social Sciences*, **338**:1–12.

Bernfeld, S. (1938). Types of adolescence. *Psychoanalytic Quarterly*, **7**:243–253.

Blaine, G. B., and McArthur, C. C. (1961). *Emotional Problems of the Student*. New York: Appleton.

Blos, P. (1962). *On Adolescence*. New York: Macmillan (Glencoe Press).

Bone, R. A. (1958). *The Negro Novel in America*. New Haven: Yale University Press.

Bordua, D. J. (1961). Delinquent subcultures. *Annals of the American Academy of Political and Social Sciences*, **338**:119–136.

Brody, E. B. (1961). "Social conflict and schizophrenic behavior in young adult negro males. *Psychiatry*, **24**:337–346.

Brody, E. B. (1963). Color and identity conflict in young boys. *Archives of General Psychiatry*, **10**:354–360.

Brody, E. (Ed.) (1968). *Minority Group Adolescents in the United States*. Baltimore: Williams and Wilkins.

Brown, C. (1965). *Manchild in the Promised Land*. New York: Macmillan.

Brown, M. C. (1955). Status of jobs and occupations as evaluated by an urban negro sample. *American Sociological Review,* **20**:561–566.

Brown, S. R. (1968). Bibliography on *Q* technique and its methodology. *Perceptual and Motor Skills,* **26**:587–613.

Budyk, J. (1961). Female adolescent development in the lower class. Unpublished undergraduate paper, Radcliffe College.

Chein, I., Gerard, D. L., Lee, R. S., and Rosenfeld, E. (1964). *The Road to H: Narcotics, Delinquency and Social Policy.* New York: Basic Books.

Clark, K. (Ed.) (1963). *The Negro Protest.* Boston: Beacon Press.

Clark, K. (1965). *Dark Ghetto: Dilemmas of Social Power.* New York: Harper.

Cleaver, E. (1968). *Soul on Ice.* New York: Delta.

Cohen, A. K. (1955). *Delinquent Boys.* New York: Macmillan (Glencoe Press).

Coleman, J. S. (1961). *The Adolescent Society.* New York: Macmillan (Glencoe Press).

Coles, R. (1963). Southern children under desegregation. *American Journal of Psychiatry,* **120**:332–344.

Coles, R. (1964). A matter of territory. *Journal of Social Issues,* **20**:43–53.

Coles, R. (1965a). Private problems and public evil: Psychiatry and segregation. *Yale Review,* **14**:513–531.

Coles, R. (1965b). Racial conflict and a child's question. *Journal of Nervous and Mental Disease,* **140**:162–170.

Coles, R. (1965c). It's the same but it's different. *Daedalus,* **94**:1107–1132.

Coles, R. (1967). *Children of Crisis.* Boston: Little, Brown.

Colley, T. (1959). Psychological sexual identity. *Psychological Review,* **66**:165–177.

Cummings, E. E. (1959). *I: Six Nonlectures.* Cambridge: Harvard University Press.

Curran, F. J., and Frosh, J. (1942). The body image in adolescent boys. *Journal of Genetic Psychology,* **60**:37–60.

Dai, B. (1955). Some problems of personality development among negro children. In Kluckhohn, Clyde, et al. (Eds.), *Personality in Nature, Society and Culture.* New York: Knopf, pp. 545–566.

Davie, M. R. (1949). *Negroes in American Society.* New York: McGraw Hill.

Davis, A., and Dollard, J. (1941). *Children of Bondage.* Washington: American Council on Education.

Davis, A., and Havighurst, R. J. (1946). Social class and color difference in child rearing. *American Sociological Review,* **11**:698–710.

Davis, K. (1958). Mental hygiene and class structure. In Stein, H., and Cloward, R. (Eds.), *Social Perspectives on Behavior.* New York: Macmillan (Glencoe Press).

Dennis, N. (1960). *Cards of Identity.* New York: Meridian.

Derbyshire, R. L., and Brody, E. (1964a). Marginality, identity and behavior in the American negro: A functional analysis. *International Journal of Social Psychiatry,* **10**:7–13.

Derbyshire, R. L., and Brody, E. (1964b). Social distance and identity conflict in negro college students. *Sociology and Social Research,* **48**:301–314.

Deutsch, H. (1967). *Selected Problems of Adolescence.* New York: International Universities Press.

Dignan, M. H. (1965). Ego identity and maternal identification. *Journal of Personality and Social Psychology,* **1**:476–483.

Dinitz, S., Scarpitti, F. R., and Reckless, W. C. (1962). Delinquency vulnerability: A cross group and longitudinal analysis. *American Sociological Review,* **27**:515–517.

Dorris, R. J., Levinson, J., and Hanfmann, E. (1954). Authoritarian personality studied by a new variation of the sentence completion test. *Journal of Abnormal and Social Psychology,* **49**:99–108.

Drake, S. C., and Cayton, H. R. (1962). *Black Metropolis: A Study of Negro Life in a Northern City.* New York: Harper.

Drake, S. C. (1965). The social and economic status of the negro in the United States. *Daedalus,* **94**:771–814.

Dreger, R. M., and Miller, K. S. (1960). Comparative psychological studies of negroes and whites in the United States. *Psychological Bulletin,* **57**:361–402.

Dreger, R. M., and Miller, K. S. (1968). Comparative psychological studies of negroes and whites in the United States. *Psychological Bulletin.* **70**:1.

Dubois, C. (1944). *The People of Alor.* Minneapolis: University of Minnesota Press.

Dubois, W. E. (1903). *The Souls of Black Folk: Essays and Sketches.* Chicago: McClurg.

Elkin, F., and Westley, W. A. (1955). The myth of adolescent culture. *American Sociological Review,* **20**:680–684.

Elkins, S. M. (1963). *Slavery.* New York: Grosset and Dunlap.

Ellison, R. (1952). *Invisible Man.* New York: Random House.

Encounter Editors (1963). Negro crisis. *Encounter,* **21**, August.

Epstein, R. (1963). Social class membership and early childhood memories. *Child Development,* **34**:503–508.

Erikson, E. (1950). *Childhood and Society.* New York: Norton.

Erikson, E. (1956). Ego identity and the psycho-social moratorium. In Witmer, H., et al. (Eds.), *New Perspectives on Delinquency.* Washington, D.C.: U.S. Government Printing Office.

Erikson, E., and Erikson, K. (1957). The confirmation of the delinquent. *Chicago Review,* **10**:15–23.

Erikson, E. (1958). *Young Man Luther.* New York: Norton.

Erikson, E. (1959). Identity and the life cycle. *Psychological Issues,* **1**:1–171.

Erikson, E. (1960). The roots of virtue. In Huxley, J. (Ed.), *The Humanist Frame.* New York: Harper.

Erikson, E. (1962). Youth: Fidelity and diversity. *Daedalus,* **91**:5–26.

Erikson, E. (1964). Memorandum on identity and negro youth. *Journal of Social Issues,* **20**:29–42.

Erikson, E. (1968). *Identity: Youth and Crisis.* New York: Norton.

Erikson, E. (1969). *Ghandi's Truth.* New York: Norton.

Evans, J. (1950). *Three Men.* New York: Knopf.

Fanon, F. (1962). *Black Skin, White Masks.* New York: Grove Press.

Fisher, S., et al. (1957). Body boundaries and style of life. *Journal of Abnormal and Social Psychology,* 52:373–379.

Frazier, E. F. (1939). *The Negro Family in the United States.* Chicago: University of Chicago Press.

Frazier, E. F. (1957). *Black Bourgeoisie.* New York: Macmillan (Glencoe Press).

Freedomways Editors (1965). Freedomways: A quarterly review of the negro freedom movement. *Freedomways,* 5.

Freud, A. (1948). *The Ego and the Mechanisms of Defense.* London: Hogarth Press.

Fried, M., and Lindeman, E. (1961). Socio-cultural factors in mental health. *American Journal of Orthopsychiatry,* 31:87–101.

Friedenberg, E. Z. (1960). *The Vanishing Adolescent.* Boston: Beacon Press.

Fromm, E. (1956). *The Art of Loving.* New York: Harper.

Ginzburg, E., et al. (1951). *Occupational Choice.* New York: Columbia University Press.

Gladstone, Herman P. (1962). Psychotherapeutic techniques with youthful offenders. *Psychiatry,* pp. 147–159.

Glaser, D. (1958). Dynamics of ethnic identification. *American Sociological Review,* 23:31–40.

Glenn, Norval D. (1963). Negro prestige criteria. *American Journal of Sociology,* 68:645–657.

Glicksberg, G. (1960). Psychoanalysis and the negro. *Phylon,* 21:337.

Goffman, E. (1963). *Stigma.* Englewood Cliffs, N.J.: Prentice-Hall.

Goodman, P. (1960). *Growing Up Absurd.* New York: Random House.

Graubard, S. R. (1962). *Youth: Change and Challenge. Daedalus,* 91:1–239.

Greenacre, P. (1958). Early physical determinants in the development of a sense of identity. *Journal of American Psychoanalytic Association,* 6:612–627.

Grier, W. H., and Cobb, P. M. (1968). *Black Rage.* New York: Basic Books.

Griffin, J. H. (1961). *Black Like Me.* Boston: Houghton Mifflin.

Hansberry, L. (1959). *A Raisin in the Sun.* New York: Random House.

Hartmann, H. (1939). *Ego Psychology and the Problem of Adaptation.* New York: International Universities Press.

Hartmann, H., and Kris, E. (1945). The genetic approach in psychoanalysis. *The Psychoanalytic Study of the Child,* 1:11–30. New York: International Universities Press.

Hartmann, H., Kris, E., and Lowenstein, R. M. (1949). Some psychoanalytic comments on culture and personality. In Wilbur, G. (Ed.), *Psychoanalysis and Culture.* New York: International Universities Press.

Hauser, S. T. (1962). Patterns of estrangement. Unpublished manuscript, Harvard University.

Hauser, S. T. (1966). *Racial and Social Contexts of Ego Identity.* M.D. Thesis, Yale University School of Medicine.

Havighurst, R. J., Bowman, P. H., Matthews, C. V., and Pierce, J. V. (1962). *Growing Up in River City.* New York: Wiley.

Havighurst, R. J., and Taba, H. (1949). *Adolescent Character and Personality.* New York: Wiley.

Hearn, R. (1963). Notes on negro life in a New Haven neighborhood. Unpublished mimeo.

Hernton, C. C. (1965). *Sex and Racism in America.* Garden City: Doubleday.

Himes, J. S. Negro teen age culture. *Annals of the American Academy of Political and Social Science,* **338**:91–101.

Hollingshead, A. (1949). *Elmtown's Youth.* New York: Wiley.

Hollingshead, A. (1957). Two factor index of social class. New Haven: Yale University, mimeo.

Hollingshead, A., and Redlich, F. C. (1958). *Social Class and Mental Illness: A Community Study.* New York: Wiley.

Howard, L. P. (1960). Identity conflicts in adolescent girls. *Smith College Studies in Social Work,* **31**:1–21.

Hyman, H. H. (1959). Value systems of different classes. In Stein, H., and Cloward, R. (Eds.), *Social Perspectives on Behavior.* New York: Macmillan (Glencoe Press), pp. 315–330.

Isaacs, H. R. (1962). *The New World of Negro Americans.* New York: J. Day.

Jacobson, E. (1964). *The Self and the Object World.* New York: International Universities Press.

Jones, L. (1964). *Dutchman and The Slave, Two Plays.* New York: Morrow.

Jones, L. (1967). *Tales.* New York: Grove Press.

Kaplan, B. (Ed.) (1961). *Studying Personality Cross-Culturally.* Evanston, Ill.: Row, Peterson.

Kardiner, A., and Ovessey, L. (1951). *The Mark of Oppression.* New York: Norton.

Karon, B. P. (1958). *The Negro Personality.* New York: Springer.

Keller, S. (1963). The social world of the urban slum child: Some early findings. *American Journal of Orthopsychiatry,* **32**:823–831.

Keniston, K., and Scott, P. (1959). Exploratory research on identity. Mimeo.

Keniston, K. (1961). Alienation and the decline of Utopia. *American Scholar,* **29**:161–200.

Keniston, K. (1962). Social change and youth in America. *Daedalus,* **91**:145–171.

Keniston, K. (1966). *The Uncommitted: Alienated Youth in American Society.* New York: Harcourt.

Klapp, O. (1962). *Heroes, Fools and Villains.* Englewood Cliffs, N.J.: Spectrum, Prentice-Hall.

Klausner, S. Z. (1964). Social class and self-concept. *Journal of Social Psychology,* **38**:201–205.

Korbin, S. (1961). Sociological aspects of the development of a street corner group. *American Journal of Orthopsychiatry,* **31**:685–702.

Kohn, M. L. (1963). Social class and parent-child relationships: An interpretation. *American Journal of Sociology,* **68**:471–480.

Korbin, S. (1962). Impact of cultural factors on the selected problems of adolescent development in the middle and lower class. *American Journal of Orthopsychiatry,* **32**:387–390.

Lederer, W. (1964). Dragons, delinquents and destiny. *Psychological Monographs* #15, pp. 1–80.

Leighton, A. H. (1959). *My Name is Legion.* New York: Basic Books.

Levitt, M., and Rubenstein, B. (1964). Some observations on the relationship between cultured variants and emotional disorders. *American Journal of Orthopsychiatry,* **34**:423–432.

Lichenstein, H. (1961). Identity and sexuality: A study of their interrelationships in man. *Journal of the American Psychoanalytic Association,* **9**:179–260.

Liebow, E. (1967). *Tally's Corner.* Boston: Little, Brown.

Lifton, R. J. (1961). *Thought Reform and the Psychology of Totalism.* New York: Norton.

Lifton, R. J. (1962). Youth and history: Individual change in post war Japan. *Daedalus,* **91**:172–197.

Lincoln, C. E. (1962). *The Black Muslims in America.* Boston: Beacon Press.

Lomas, P. (1961). Family role and identity formation. *International Journal of Psychoanalysis.* **42**:371–380.

Lomax, L. E. (1962). *The Negro Revolt.* New York: Harper.

Lomax, L. E. (1963). *When the Word is Given.* New York: World.

Lott, A., and Lott, B. E. (1963). *Negro and White Youth: A Psychological Study in a Border State Community.* New York: Holt.

Lynd, H. M. (1958). *On Shame and the Search for Identity.* New York: Harcourt.

McDaniel, P. A., and Babchuk, N. (1960). Negro conceptions of white people in a northeastern city. *Phylon,* **21**:7–19.

Machover, K. (1949). *Personality Projection in the Drawing of the Human Figure.* Springfield, Ill.: Charles C Thomas.

MacInnes, C. (1963). Dark angel. *Encounter,* **21**, August.

McNemar, Q. (1955). *Psychological Statistics.* New York: Wiley.

Malcolm X (1968). *The Autobiography of Malcolm X.* New York: Grove Press.

Marcia, J. E. (1966). Development and validation of ego identity status. *Journal of Abnormal and Social Psychology,* **3**:551–558.

Marcia, J. E. (1967). Ego identity status: Relationship to change in self-esteem, 'general maladjustment' and authoritarianism. *Journal of Personality,* **35**:119–133.

Martin, D. G. (1969). Consistency of self-descriptions under different role sets in neurotic and normal adults and adolescents. *Journal of Abnormal Psychology,* **74**:2, 173–176.

Miller, W. (1959). *The Cool World.* New York: Little, Brown.

Miller, W.B. (1958). Lower class culture as a generating milieu of gang delinquency. *Journal of Social Issues,* **14**:5–19.

Moynihan, D. P. (1965). Employment, income, and the ordeal of the negro family. *Daedalus,* **94**:745–770.

Murray, H. A. (1938). *Explorations in Personality.* London: Oxford University Press.

Mussen, P. H. (1953). Differences between the TAT responses of negro and white boys. *Journal of Consulting Psychology,* **17**:373–376.

Mussen, P. H., and Jones, M. C. (1959). Self concepts and motivations of early and late developing adolescents. *Child Development,* **28**:243–256.

Myers, J. K., and Roberts, B. H. (1959). *Family and Class Dynamics in Mental Illness.* New York: Wiley.

Myrdal, G. (1944). *An American Dilemma.* New York: Harper.

Newsweek Editors (1963). The Negro in America. *Newsweek,* July 29, 1963.

Nunberg, H. (1931). The synthetic function of the ego. *International Journal of Psychoanalysis,* **12**:123–140.

Otnow, D., and Prelinger, E. (1962). An abstract design test of the capacity for intimacy. *Perceptual and Motor Skills,* **15**:645–647.

Parsons, T. (1965). Full citizenship for the American negro: A sociological problem. *Daedalus,* **94**:1009–1054.

Pettigrew, T. F. (1964a). Negro American personality: Why isn't more known? *Journal of Social Issues,* **20**:4–23.

Pettigrew, T. F. (1964b). *A Profile of the American Negro.* New York: Van Nostrand Reinhold.

Pettigrew, T. F. (1964c). Race, mental illness and intelligence. *Eugenics Quarterly,* **11**:189–215.

Pettigrew, T. F. (1965). Complexity and change in American racial patterns: A sociological view. *Daedalus,* **94**:974–1008.

Phillips, U. B. (1918). *American Negro Slavery: A Survey of the Supply, Employment and Control of Negro Labor as Determined by the Plantation Regime.* New York: Appleton.

Phylon Editors (1964). *Phylon: The Atlantic University Review of Race and Culture,* **25**.

Pierce, C. (1968). Problems of the negro adolescent in the next decade. In Brody, E. (Ed.), *Minority Group Adolescents in the United States.* Baltimore: Williams and Wilkins.

Piers, G., and Singer, M. B. (1953). *Shame and Guilt.* Springfield, Ill.: Charles C Thomas.

Polsky, H. W., and Kohn, M. (1959). Participant observation in a delinquent subculture. *American Journal of Orthopsychiatry,* **29**:737–751.

Pouissant, A. (1967). Negro self-hate. *New York Times Sunday Magazine,* August 20, 1967.

Pouissant, A., and Ladner, J. (1968). Black power. *Archives of General Psychiatry,* **18**:385–398.

Powdermaker, H. (1943). The channeling of negro aggression by the cultural process. *American Journal of Sociology,* **48**:750–758.

Prelinger, E. (1958). Identity diffusion and the synthetic function. In Wedge, B.

M. (Ed.), *Psychosocial Problems of College Men.* New Haven: Yale University Press.

Prelinger, E. (1960). Discussion of papers by Dr. E. Slocombe and Dr. J. Wilms on problems of ego identity in college students. Unpublished manuscript.

Prelinger, E., and Zimet, C. N. (1964). *An Ego Psychological Approach to Character Assessment.* Macmillan (Glencoe Press).

Prelinger, E., Zimet, C., and Levin, M. (1960). An ego psychological scheme for personality assessment. *Psychological Reports,* **7**:182f.

Rainwater, L. (1966). Crucible of identity: The negro lowerclass family. *Daedalus,* **95**:172–216.

Rainwater, L., and Yancey, W. (Eds.) (1967). *The Moynihan Report and the Politics of Controversy.* Cambridge: M.I.T. Press.

Rappaport, D. (1959). A historical survey of psychoanalytic ego psychology. *Psychological Issues,* **1**:5–17.

Redding, S. (1951). *On Being Negro in America.* Indianapolis: Bobbs-Merrill.

Riesman, D. (1950). *The Lonely Crowd.* New Haven: Yale University Press.

Rohrer, J. H., and Edmonson, M. S. (Eds.) (1960). *The Eighth Generation.* New York: Harper.

Rose, A. M. (1956). Psychoneurotic breakdown among soldiers in combat. *Phylon,* **17**:61–69.

Rose, A. M. (1965). The negro protest. *Annals of the American Academy of Political and Social Science,* #357.

Rosenberg, M. (1965). *Society and the Adolescent Self Image.* Princeton, Princeton University Press.

Ross, A. O. (1962). Ego identity and the social order. *Psychological Monographs,* #542.

Sanford, N. (Ed.) (1962). *The American College.* New York: Wiley.

Schachtel, E. G. (1962). On alienated concepts of identity. *Journal of Humanist Psychology,* **1**:110–121.

Schafer, R. (1968). *Aspects of Internalization.* New York: International Universities Press.

Schatzman, L., and Strauss, A. L. (1954). Social class and modes of communication. *American Journal of Sociology,* **60**:329–338.

Scher, M. (1967). Negro group dynamics. *Archives of General Psychiatry,* **17**: 646–651.

Schilder, P. (1951). *The Image and Appearance of the Human Body.* New York: International Universities Press.

Sclare, A. B. (1953). Cultural determinants in the neurotic negro. *British Journal of Medical Psychology,* **26**:278–288.

Shapiro, R. (1963). Adolescence and the psychology of the ego. *Psychiatry,* **26**:77–87.

Short, J. F., and Strodtbeck, F. L. (1963). Response of gang leaders to status threats. *American Journal of Sociology,* **68**:571–579.

Siegel, S. (1956). *Nonparametric Statistics.* New York: McGraw-Hill.

Silberman, C. E. (1964). *Crisis in Black and White.* New York: Random House.

Sklarew, B. H. (1959). Effects on adolescents of different sexes of early separation from parents. *Psychiatry,* **22**:399–405.

Spiegel, L. (1951). A review of the contributions to a psychoanalytic theory of adolescence: Individual aspects. *The Psychoanalytic Study of the Child,* **6**:375–393.

Spiegel, L. (1958). Comments on the psychoanalytic psychology of adolescence. *The Psychoanalytic Study of the Child,* **13**:296–308.

Srole, L., Langner, T. S., Michael, S. T., Opler, M., and Rennie, T. (1962). *Mental Health in the Metropolis: The Midtown Manhattan Study,* Vol. I. New York: McGraw-Hill.

Stein, H., and Cloward, R. (Eds.) (1958). *Social Perspectives on Behavior.* New York: Macmillan (Glencoe Press).

Stephenson, W. (1953). *The Study of Behavior.* Chicago: University of Chicago Press.

Strauss, A. L. (1959). *Mirrors and Masks: The Search for Identity.* New York: Macmillan (Glencoe Press).

Sullivan, H. S. (1953). *The Interpersonal Theory of Psychiatry.* New York: Norton.

Symonds, P. N., and Jensen, A. R. (1961). *From Adolescent to Adult.* New York: Columbia University Press.

Weaver, E. K. (1956). Racial sensitivity among negro children. *Phylon,* **17**:52–60.

Wedge, B. M. (Ed.) (1958). *Psychosocial Problems of College Men.* New Haven: Yale University Press.

Wender, P. H. (1968). Vicious and virtuous circles: The role of deviation amplifying feedback in the origin and precipitation of behavior. *Psychiatry,* **31**:309–324.

Wheelis, A. (1958). *The Quest for Identity.* New York: Norton.

White, R. K. (1947). Black boy: A value analysis. *Journal of Abnormal and Social Psychology,* **42**:440–461.

White, R. W. (1952). *Lives in Progress.* New York: Dryden Press.

White, R. W. (1963). Sense of interpersonal competence: Two case studies and some reflections on origins. In White, R. W., *The Study of Lives.* Englewood Cliffs, N.J.: Prentice-Hall.

White, R. W. (1963). *The Study of Lives.* New York: Atherton.

Whyte, W. F. (1955). *Street Corner Society.* Chicago: University of Chicago Press.

Whyte, W. F. (1958). A slum sex code. In Stein, H., and Cloward, R. (Eds.), *Social Perspectives on Behavior.* New York: Macmillan (Glencoe Press), pp. 441–448.

Witmer, H. (1956). *New Perspectives in Delinquency.* Washington, D.C.: U.S. Government Printing Office.

Wildstad, A. M. (1961). Identity conflicts in disturbed adolescent girls. *Smith College Studies in Social Work,* **32**:20–37.

Wright, R. (1945). *Black Boy*. New York: Harper.

Wylie, R. C. (1961). *The Self Concept: A Critical Survey of Pertinent Research Literature*. Lincoln: University of Nebraska Press.

Young, F. M. (1959). Response of juvenile delinquents to the TAT. *Journal of Genetic Psychology*, **88**:251–259.

Name Index

Subject Index

Date Due

JY 5